The Bible and the
Age of the Earth

Bert Thompson, Ph.D.

APOLOGETICS PRESS

Apologetics Press, Inc.
230 Landmark Drive
Montgomery, Alabama 36117-2752

TABLE OF CONTENTS

1

INTRODUCTION

In the current controversy over creation and evolution, it is a rare event indeed to find something on which those in both camps agree wholeheartedly. Generally speaking, the two world views are light-years apart from start to finish. There is one thing, however, on which both creationists and evolutionists **do** agree: evolution is impossible if the Earth is young (with an age measured in thousands, not billions, of years). R.L. Wysong addressed this point in his book, *The Creation-Evolution Controversy*.

> Both evolutionists and creationists believe evolution is an impossibility if the universe is only a few thousand years old. There probably is no statement that could be made on the topic of origins which would meet with so much agreement from both sides. Setting aside the question of whether vast time is competent to propel evolution, we must query if vast time is indeed available (1976, p. 144).

It may be somewhat ironic that so much discussion has resulted from something on which both sides seemingly agree, but it should not be at all surprising. Apart from the most basic issue of the controversy itself—i.e., whether creation or evolution is the correct view of origins—the single most serious area of conflict between those who accept the biblical account of creation and those who accept the theory of organic evolution (in whole or in part) is the chronological framework of history—viz., the age of the Earth. And, of course, this subject is of intense interest not only to those who promulgate atheistic evolution, but to those who are sympathetic with certain portions of that theory as well. While a young Earth/Universe presents no problem at all for creationists who accept the biblical account of origins at face value, it is the death knell to almost every variety of the evolutionary scenario.

A simple, straightforward reading of the biblical record indicates that the Cosmos was created in six days only a few thousand years ago. Standing in stern opposition to that view is the suggestion of atheistic evolutionists, theistic evolutionists, progressive creationists, and so-called "old-Earth creationists" that the current age of the Universe can be set at roughly 8-12 billion years, and that the Earth itself is almost 5 billion years old. Further complicating matters is the fact that the biblical record plainly indicates that living things were placed on the newly created Earth even before the end of the six-day creative process (e.g., plant life came on day three). The evolutionary scenario, however, postulates that primitive life evolved from nonliving chemicals roughly 3.5-4.0 billion years ago and that all other life forms gradually developed during the alleged "geologic ages" (with man arriving on the scene, in one form or another, approximately 1-2 million years ago).

Even to a casual observer, it is apparent that the time difference involved in the two models of origins is significant. Much of the controversy today between creationists, atheistic evolutionists, theistic evolutionists, progressive creationists, and old-Earth creationists centers on the age of the Earth. The magnitude of the controversy is multiplied by three factors. First, atheistic evolution itself is impossible to defend if the Earth is young. Second, the concepts mentioned above that are its "theistic cousins" likewise are impossible to defend if the Bible is correct in its straightforward teachings and obvious implications about the age of the Earth. Third, there is no possible compromise that will permit the old-Earth/young-Earth scenarios to coexist; the gulf separating the biblical and evolutionary views in this particular area simply is too large. As Henry Morris correctly observed:

> Thus the Biblical chronology is about a million times shorter than the evolutionary chronology. A million-fold mistake is no small matter, and Biblical scholars surely need to give primary attention to resolving this tremendous discrepancy right at the very foundation of our entire Biblical cosmology. This is not a peripheral issue that can be dismissed with some exegetical twist, but is central to the very integrity of scriptural theology (1984, p. 115).

In the earlier quote from Dr. Wysong, it was suggested that we must "query if vast time is indeed available." That is exactly what this book intends to do. Indeed, a million-fold mistake is no small matter. How old is the Earth according to God's Word?

THE AGE OF THE EARTH—"WAIT AND SEE"

As I begin this investigation into the age of the Earth, I first would like to define the scope of the present inquiry. The title of this book is *The **Bible** and the Age of the Earth*. It is not my intention here to examine and/or refute the scientific evidences that allegedly establish an ancient Earth. There are a number of books available that provide such information (see, for example: Ackerman, 1986; Henry Morris, 1974, 1989; Jackson, 1989; Kautz, 1988; John Morris, 1994; Morris and Parker, 1987; Woodmorappe, 1999; Wysong; 1976). Rather, I intend to limit my discussion to what God's Word has to say on this subject.

Obviously, then, I am not writing with the atheistic evolutionist in mind. I am well aware that my arguments would carry no weight whatsoever with the person who falls into that specific category. Rather, this discussion is intended for those who: (a) believe in the God of the Bible; (b) claim to accept the Bible as His inspired, authoritative Word; and (c) are convinced that what God has said can be understood. For such a person, the Bible is the recognized, final authority on any subject that it addresses. Renowned biblical scholar Edward J. Young expressed this point well when he wrote:

> It is of course true that the Bible is not a textbook of science, but all too often, it would seem, this fact is made a pretext for treating lightly the content of Genesis one. Inasmuch as the Bible is the Word of God, whenever it speaks on any subject, whatever that subject may be, it is accurate in what it says (1964, p. 43).

The question then becomes: "**Does** the Bible address the age of the Earth?" Yes, it does. But before we delve into what it says, there are two popular, prevailing attitudes that need to be discussed.

First, I acknowledge that some religionists regard this as a question that simply cannot be answered at present. We are urged to "wait and see" or to "reserve judgment." Jack Wood Sears, former chairman of the biology department at Harding University in Searcy, Arkansas, wrote:

> When conflicts do occur, the part of wisdom is to withhold judgment until the facts are all in. For example, there is difficulty with the age of life on the earth. Science, as I indicated earlier, has seemed to indicate that the life has been here much longer than we have generally interpreted the Bible to indicate. However, scientific determination of the ages of geological strata is not absolute and is subject to much difficulty and uncertainty. The Bible, as we have shown, does not date creation, **and the intimations it seems to present** may not be properly understood. Since I hold science to be a valid approach to reality, and since I have concluded upon much and sufficient evidence, that the Bible is inspired and therefore true, the only rational recourse, it seems to me, is to withhold judgment about a seeming contradiction. **Wait and see** (1969, p. 97, emp. added).

Four years later, J. Frank Cassel wrote in a similar vein.

> The thoughtful person respects present knowledge in both areas (science and Biblical research) and keeps searching for new information and insight. In the meantime he must **reserve judgment**, saying simply "I don't know where the proper synthesis lies." The tension remains as the search continues (1973, pp. 251-252, emp. added).

While at first glance such an attitude may appear to be laudable, I would like to suggest that it is nothing but a ruse. Authors of such sentiments no doubt want **others** to adhere to their advice, but they themselves have absolutely no intention of doing so.

Cassel, for example, has written often about the accuracy of the geologic timetable and is a well-known apologist for the old-Earth world view. Further, in November 1983 I debated Dr. Sears on the topic of the age of the Earth.* I affirmed the proposition that the Bible does **not** allow for an ancient Earth; Dr. Sears affirmed the proposition that it **does**. The debate occurred 14 years **after** Dr. Sears penned his "wait and see" statement. Had he discovered additional information during those years that no longer made it necessary to wait and see? Apparently not, since during the debate he told the audience he was "still waiting" (an exact quote from the transcript) for information that would allow him to make a decision about the age of the Earth. If he was still waiting, **why, then, would he then be so willing to engage in a public debate to defend the proposition that the Bible allows for an ancient Earth**? Where is the consistency in such a position?

In reality, what these writers mean when they say that "we" should "wait and see," or that "we" should "reserve judgment" is that **those who believe in a young Earth** should wait and see or reserve judgment. In the meantime, **they** will continue to advocate publicly their position that an ancient Earth is wholly consistent with the biblical record.

---

* The debate is available in printed, audio, and video formats. The printed manuscripts of the debate are in McClish (1983), pp. 405-434. Audio and video tapes are available from Apologetics Press.

Second, there are some in the religious community who suggest that the Bible is conspicuously silent on the topic of the age of the Earth. It is not uncommon to hear statements suggesting that since the Bible does not address this matter, a person is free to believe whatever he or she wishes in this regard. Typical of such a mind-set are these statements by Donald England and John Clayton.

> However, nowhere does a Biblical writer give us an age for earth or an age for life on earth.... Inasmuch as Scripture does not state how old the earth is or how long life has existed on earth, one is free to accept, if he wishes, the conclusions of science (England, 1983, pp. 155-156).

> Genesis 1:1 is an undated verse. No time element is given and no details of what the Earth looked like are included. It could have taken place in no time at all, or God may have used eons of time to accomplish his objectives (Clayton, 1976a, pp. 147-148).

This, of course, is but another ruse. Beware when a writer or speaker suggests that the Bible is "silent" on the topic of the age of the Earth or that a person is free to accept the varied "conclusions of science." What those who make such statements **really** mean is that **they** are free to accept the conclusions, not of science, but of uniformitarian geology, and in so doing to defend the same old-Earth position as their evolutionist colleagues. Both England and Clayton, for example, are on record as defending an ancient Earth (see: England, 1972, pp. 103-106; Clayton as documented in Jackson and Thompson, 1992, pp. 99-110).

CHRONOLOGY AND THE BIBLE

The truth of the matter is that the Bible, being a book grounded in history, is filled with chronological data that may be used to establish a relative age for the Earth. It is not "silent" on this topic, and thus there is no need to "wait and see" or to "reserve judgment." Professor Edwin Thiele, the man who unlocked much of the mystery of Old Testament chronology, declared:

> We know that God regards chronology as important, for He has put so much of it into His Word. We find chronology not only in the historical books of the Bible, but also in the prophetic books, in the Gospels, and in the writings of Paul (1977, p. 7).

The Bible, for example, provides impressive chronological data from Adam to Solomon. Combining information from the Assyrian Eponym Lists and the Black Obelisk, the death of Ahab has been determined to be 853-852 B.C. (Packer, et al., 1980, p. 48) and therefore the reign of Solomon (some forty years, 1 Kings 11: 42) can be dated at 971-931 B.C. (Merrill, 1978, p. 97; Packer, et al., 1980, p. 50; Brantley, 1993, p. 83). According to 1 Kings 6:1, 480 years before Solomon's fourth year of reign (967-966 B.C.), Moses brought the Israelites out of Egypt. The date of the Exodus is 1446/1445 B.C. (Unger, 1973, pp. 140-152; Archer, 1970, pp. 212-222; Packer, et al., 1980, p. 51; Jackson, 1981, p. 38; 1990, p. 17).

To this date is added the years of sojourn in Egypt (430 years, Exodus 12:40), thereby producing the date of 1876 B.C. as the year Jacob went to Egypt (Packer, et al., 1980, p. 50). Interestingly, the Bible records Pharaoh's query of Jacob's age (and Jacob's answer—130 years) in Genesis 47:9. This would make the year of Jacob's birth 2006 B.C. (Genesis 25:26). Abraham was 100 years old when he begat Isaac, giving the date of 2166 B.C. for Abraham's birth (Genesis 21:5; Packer, et al., 1980, p. 54). The chronology from Abraham to Adam is recorded very carefully in two separate genealogical tables—Genesis 5 and 11. According to Genesis 12:4, Abraham was 75 when he left Haran, presumably after Terah died at 205 years; thus, Abraham was born when Terah was 130 years old, albeit he is mentioned first by importance when Terah started having sons at the age of 70 (Genesis 11:27; 12:4; Acts 7:4).

Having established the birth date of Abraham at 2166 B.C. (Archer, 1970, pp. 203-204), it is possible to work from the time of Adam's creation to Abraham in order to discern the chronology of "the beginning." The time from the creation of Adam to Seth was 130 years (Genesis 5:3), the time from Adam to Noah was 1056 years (Packer, et al., 1980, pp. 56-57), and the time from Noah's birth to the Flood was 600 years (Genesis 7:6), or 1656 A.A. (After Adam). It appears that Shem was about 100 years old at the time of the Flood (Genesis 5:32; 11:10) and begat Arphaxad two years after the Flood (the Earth was not dry for more than a year; cf. Genesis 7:11 with 8:14; see also Genesis 11:10) in approximately 1659 A.A.

Arphaxad begat Salah in his 35th year; however, Luke 3:36 complements the chronological table of Genesis 11 with the insertion of Cainan between Arphaxad and Salah, which indicates that likely Arphaxad was the father of Cainan. Proceeding forward, one observes that Terah was born in 1879 A.A. and bore Abraham 130 years later (in 2009 A.A.). Simple arithmetic—2166 B.C. added to 2009 A.A.—places the creation date at approximately 4175 B.C. The Flood, then, would have occurred around 2519 B.C.

Numerous objections have been leveled at the literal and consecutive chronological interpretation of Scripture. For example, some have suggested that the tables of Genesis 5 and 11 are neither literal nor consecutive. Yet five of the patriarchs clearly were the **literal** fathers of their respective sons: Adam named Seth (Genesis 4:25), Seth named Enos (4:26), Lamech named Noah (5:29), Noah's literal, consecutive sons were Shem, Ham and Japheth (cf. 5:32 with 9:18), and Terah fathered Abraham directly (11:27,31). Jude's record in the New Testament counts Enoch as "the seventh from Adam" (Jude 1:14), thereby acknowledging the genealogical tables as literal and consecutive. Moreover, how better could Moses have expressed a literal and consecutive genealogy than by using the terms "lived...and begat...begat...after he begat...all the days...and he died"? Without question, Moses noted that the first three individuals (Adam, Seth, and Enos) were consecutive, and Jude stated by inspiration that the first seven (to Enoch) were consecutive. Enoch's son, Methuselah, died the year of the Flood, and so by three steps the chronology of Adam to Noah is literal and consecutive, producing a trustworthy genealogy/chronology.

There have been those who have objected to the suggestion that God is concerned with providing information on the age of the Earth and humanity. But the numerous chronological tables permeating the Bible prove that theirs is a groundless objection. God, it seems, **was very concerned** about giving man exact chronological data and, in fact, was so concerned that He provided a precise knowledge of the period back to Abraham, plus two tables—with ages—from Abraham to Adam. The ancient Jewish historians (1 Chronicles 1:1-27) and the New Testament writers (Luke 3:34-48) understood the tables of Genesis 5 and 11 as literal and consecutive. The Bible explains quite explicitly that God created the Sun and Moon to be timekeepers (Genesis 1:16) for Adam and his descendants (notice how Noah logged the beginning and the ending of the Flood using these timekeepers, Genesis 6:11; 9:14).

Still others have suggested that the two tables somehow are symbolic. But the use (or even repetitive use) of a "unique" number does not necessitate a symbolical interpretation. Special numbers (such as 7,10,12,40, etc.) employed in Scripture may be understood as literal despite the frequency of their use. Are there not three **literal** members of the Godhead? Did not Sceva have seven **literal** sons? Were there not ten **literal** commandments? Were there not twelve **literal** apostles? Was Christ's fast in the wilderness not forty **literal** days? Moreover, those who study history routinely recognize that it abounds with numerical "coincidences." To say that the tables of Genesis 5 and 11 are "symbolic" of long periods of time flies in the face of the remainder of the biblical record.

Those who believe that the Bible is **unconcerned** with chronology would do well to spend some time studying the lineages of the Hebrew kings in the Old Testament. James Jordan has explained:

> Chronology **is** of concern to the writers of the Bible. From this perspective we should be **surprised** if the Bible did not include chronological data regarding the period from Creation to Abraham, especially since such data can now be obtained from no other source. That chronology is of concern to the Bible (and to its Author) can also be seen from the often difficult and confusing chronology of the Kings of Israel. Thus, we find that it is the intention of the Bible to provide us with chronology from Abraham to the Exile. Some of that chronology is given in summary statements ...but some is also given interspersed in the histories of the Kings. Is it therefore surprising or unreasonable that some should be given along with genealogies as well? (1979/1980, p. 21, emp. and parenthetical item in orig.).

While it is true that genealogies (and chronologies) serve various functions in Scripture, one of their main purposes is to show the historical connection of great men to the unfolding of Jehovah's redemptive plan. These lists, therefore, are a link from the earliest days of humanity to the completion of God's salvation system. **In order to have any evidential value, they must be substantially complete.**

For example, the inspired writer of Hebrews, in contending for the heavenly nature of Christ's priesthood, argued that the Savior could not have functioned as a priest while He was living upon the Earth since God had in place a **levitical** priesthood to accomplish that need (Hebrews 8:4). Jesus did not qualify for the levitical priesthood because "it is **evident** that our Lord hath sprung out of **Judah**" (Hebrews 7:14, emp. added). How could it have been "evident" that Jesus Christ was from the tribe of Judah—**unless there were accurate genealogical records by which such a statement could be verified**? The writer of Hebrews based his argument on the fact that the various readers of his epistle would not be able to dispute the ancestry of Christ due to the reliable nature of the Jewish documentation available—i.e., the genealogies.

It has been argued that secular history is considerably older than 4000 B.C. But ponder this. When the studies of various Egyptologists are examined, no two give the same time period for the Old Kingdom (III-VI Dynasties). Aling (1981) dated it at 2800-2200 B.C. Baikie (1929) dated the period as 3190-2631 B.C. Breasted (1912) gave the date as 2980-2475 B.C. White (1970) suggested 2778-2300 B.C. With such variability in the last "sure" period of Egypt's history, how can dogmatism prevail for the **pre**dynastic period? Scientists and historians influence Christendom with their "established limits" of history. Theologians influence Christianity with evolution-based bias as well. For instance, Gleason Archer has stated:

> The problems attending this method of computation are compounded by the quite conclusive archaeological evidence that Egyptian Dynasty I went back to 3100 B.C., with a long period of divided kingdoms in the Nile valley before that. These could hardly have arisen until long after the Flood had occurred and the human race had multiplied considerably (cf. Genesis 10). It therefore seems necessary to interpret the figures of Genesis 5 and 11 differently, especially in view of the gaps in other biblical genealogical tables (1979, 1:361).

Obviously Archer is completely willing to override Scripture with the "scientific" message of archaeology. This mind-set—which requires the Bible to submit to science (geology, paleontology, etc.) —undermines the authority of the Word of God. In one prominent example from a few years back, the then-editor of *Christianity Today* stated:

> But one fact is clear: the genealogies of Genesis will not permit us to set any exact limit on the age of man. Of that we must remain ignorant **unless the sciences of geol-**

ogy and historical anthropology give us data from which we may draw tentative scientific conclusions (Kantzer, 1982, p. 25, emp. added).

The fact of the matter is that **both** scientists **and** theologians should be concerned with fitting the scientific data to the truth—God's Word—not with molding God's Word to fit current scientific theories (which, in a few short years may change—e.g., in Charles Darwin's day, the Earth had been "proven" scientifically to be 20 million years old, while today it has been "proven" scientifically to be 4.6 billion years old).

Furthermore, archaeologists often use speculative (and inaccurate) techniques such as radiocarbon dating, dendrochronology (tree-ring analysis), and pottery dating schemes. Yet each of these methods is beset with serious flaws, not the least of which are the basic assumptions upon which they are constructed. In two timely, well-researched articles ("Dating in Archaeology: Radiocarbon & Tree-Ring Dating," and "Dating in Archaeology: Challenges to Biblical Credibility"), Trevor J. Major (1993, 13:74-77) and Garry K. Brantley (1993, 13:81-85) explained the workings of these various methods and exposed the faulty assumptions upon which each is based. After listing and discussing five important problem areas associated with carbon-14 dating, and after discussing the problems associated with obtaining accurate tree-ring growth rates, Major wrote:

> Radiocarbon dating assumes that the carbon-12/carbon-14 ratio has stayed the same for at least the last hundred thousand years or so. However, the difference between production and decay rates, and the systematic discrepancy between radiocarbon and tree-ring dates, refute this assumption.... Similarly, we should not accept the claims for dendrochronology at face value. Bristlecones may add more than one growth ring per year, and the "art" of cross dating living and dead trees may be a considerable source of error. Both radiocarbon dating and dendrochronology face technical problems, and are loaded with old Earth ideas. They assume

that nature works today the same as it has worked for millions of years, yet the facts do not support this contention. **Neither method should give us cause to abandon the facts of biblical history** (1993, 13:77, emp. added).

In his article, Brantley addressed the problems associated with subjectivism in archaeological chronology in general and pottery dating in particular. He then drew the following conclusions:

...we must recognize that archaeological evidence is fragmentary and, therefore, greatly limited. Despite the amount of potsherds, bones, ornaments, or tools collected from a given site, the evidence reflects only a paltry fraction of what existed in antiquity (Brandfon, 1988, 14[1]:54). Unearthed data often are insufficient, inconclusive, and subject to biased interpretation....

...the paucity of archaeological evidence provides fertile soil for imaginative—and often contradictory—conclusions. We must not overlook the matter of subjectivity in interpretations.... Finally, **archaeology is an imprecise science, and should not serve as the judge of biblical historicity**. The pottery dating scheme, for example, has proved to be most helpful in determining relative dates in a tell. But, at best, pottery can place one only within the "chronological ball park." John Laughlin, a seasoned archaeologist, recognized the importance of potsherds in dating strata, but offered two warnings: (1) a standard pottery type might have had many variants; and (2) similar ceramic types might not date to the same era—some types may have survived longer than others, and different manufacturing techniques and styles might have been introduced at different times in different locales. Further, he mentioned the fact of subjectivity in determining pottery: "...in addition to its observable traits, pottery has a 'feel' to it" (1992, 18[5]:72). **Therefore, we must recognize archaeology for what it is—an inexact science with the innate capacity for mistakes** (1993, 13:84-85, emp. added).

Wayne Jackson accurately summarized the importance of biblical chronology when he observed:

The purpose of biblical chronology is to determine the correct dates of events and persons recorded in the Bible as accurately as possible, in order that we may better understand their role in the great plan of Jehovah.... The Bible is the inspired Word of God (II Tim. 3:16). Its testimony is, therefore, always reliable. Whenever it speaks with reference to chronological matters, one may be sure that it is **right**! No chronology is thus to be trusted which contradicts plain historical/chronological data in the sacred text, or which requires a manipulation of factual Bible information (such as is frequently done by compromisers who have been romanced by the chronological absurdities of the theory of evolution) [1981, 1:37, emp. and parenthetical items in orig.].

Was chronology of importance to the biblical writers? Indeed it was. Does the Bible speak, then, in any sense, concerning the age of the Earth or the age of humanity on the Earth? Indeed it does. I am not suggesting, of course, that one can settle on an **exact** date for the age of the Earth (as did John Lightfoot [1602-1675], the famed Hebraist and vice-chancellor of Cambridge University who taught that creation occurred the week of October 18 to 24, 4004 B.C., and that Adam and Eve were created on October 23 at 9:00 A.M., forty-fifth meridian time [see Ramm, 1954, p. 121]). I do contend, however, that the Bible gives a chronological framework that establishes a **relative** age for the Earth—an age confined to a span of only a few thousand years. The material that follows presents the evidence to support such a conclusion.

WHY DO WE NEED AN OLD EARTH?

In his book, *Creation or Evolution?*, D.D. Riegle observed: "It is amazing that men will accept long, complicated, imaginative theories and reject the truth given to Moses by the Creator Himself" (1962, p. 24). Why is this the case? Even proponents of the old-Earth view admit that a simple, straightforward reading of the biblical text "seems to present" a young Earth. Jack

Wood Sears, quoted earlier, has admitted concerning the biblical record that "the intimations **it seems to present** may not be properly understood" (1969, p. 97, emp. added). These "intimations" of a young Earth have not escaped those who opt for an old Earth. In 1972, Donald England, distinguished professor of chemistry at Harding University in Searcy, Arkansas, wrote in *A Christian View of Origins:*

> But why do some people insist that the earth is relatively recent in origin? First, I feel that it is because **one gets the general impression from the Bible that the earth is young.** ... It is true that Biblical chronology leaves one with the general impression of a relatively recent origin for man... (p. 109, emp. added).

Eleven years later, when Dr. England authored his book, *A Scientist Examines Faith and Evidence,* apparently his views had not changed.

> A reading of the first few chapters of Genesis **leaves one with the very definite general impression that life has existed on earth for, at the most, a few thousand years** (1983, p. 155, emp. added).

Both Sears and England admit that the Bible "intimates" a young Earth, and that a reading of the first chapters of Genesis "leaves one with the general impression" of a young Earth. Do these two men then accept a youthful planet? They do not. Why? If a simple, plain, straightforward reading of the biblical text indicates a young Earth, what reason(s) do they give for not accepting what the Bible says? Here is Dr. England's 1983 quotation again, but this time reproduced with his introductory and concluding statements:

> Third, it is not recommended that one should allow a general impression gained from the reading of Scripture to crystallize in his mind as absolute revealed truth. A reading of the first few chapters of Genesis leaves one with the very definite impression that life has existed on earth for, at the

most, a few thousand years. **That conclusion is in conflict with the conclusions of modern science that the earth is ancient** (1983, p. 155, emp. added).

In his 1972 volume, England had stated: "**From the many scientific dating methods** one gets the very strong general impression that the earth is quite ancient" (p. 103, emp. added). Dr. Sears wrote: "**Science**, as I indicated earlier, **has seemed to indicate** that life has been here much longer than we have generally interpreted the Bible to indicate" (1969, p. 97, emp. added). The professors' point, explained in detail in their writings, is this: **uniformitarian dating methods take precedence over the Bible!** Thus, scientific theory has become the father of biblical exegesis. The decisive factor no longer is "What does the Bible say?," but rather, "What do evolutionary dating methods indicate?" In order to force the biblical record to accommodate geologic time, defenders of these dating methods do indeed find it necessary to invent "long, complicated, and imaginative" theories.

One of the most important questions, then, in the controversy over the age of the Earth is this: If the Earth is ancient, **where in the biblical record** can the time be placed to guarantee such antiquity? There are but three options. The time needed to ensure an old Earth might be placed: (a) **during** the creation week; (b) **before** the creation week; or (c) **after** the creation week. If the time cannot be inserted successfully into one of these three places, then it quickly becomes obvious that an old-Earth view is unscrip-

2

THE DAY-AGE THEORY

The attempt to place the eons of time necessary for an ancient Earth **during** the creation week generally is known as the Day-Age Theory—a view which suggests that the days of Genesis 1 were not literal, 24-hour days, but rather lengthy periods or eons of time. Arthur F. Williams has observed:

> There are certain areas of biblical interpretation in which Christians find themselves in serious disagreement. One of these is the Genesis account of creation. Some interpret the record **literally**, believing each of the six days to have been cycles of 24 hours, on the sixth of which God created man in His own image by divine fiat from the dust of the earth. They believe that God breathed into man's nostrils the breath of life and he became a living soul. They, likewise, believe that this occurred at a time not longer than a few thousand years ago. Others interpret the entire record of creation "parabolically," and insist that the six days represent a vast period of time, extending into millions or billions of years (1970, p. 24, emp. in orig.).

Surburg has noted:

> Another group of interpreters has adopted what is known as the "concordistic theory." They say that the "days" of

Genesis possibly are periods of time extending over millions of years. They believe that this interpretation can be made to correspond to the various geological periods or ages. This is sometimes referred to as the "day-age" theory (1959, p. 57).

John Klotz addressed this point in *Genes, Genesis, and Evolution*: "It is hardly conceivable that anyone would question the interpretation of these as ordinary days were it not for the fact that people are attempting to reconcile Genesis and evolution" (1955, p. 87). Guy N. Woods concluded: "The day-age theory is a consequence of the evolutionary theory. But for that speculative view such a hypothesis would never have been advanced" (1976, p. 17).

IS THE DAY-AGE THEORY POPULAR?

Is the Day-Age Theory popular? Yes, and it has been advocated by a number of influential people in the religious community.

> Many sincere and competent Biblical scholars have felt it so mandatory to accept the geological age system that they have prematurely settled on the so-called day-age theory as the recommended interpretation of Genesis 1. By this device, they seek more or less to equate the days of creation with the ages of evolutionary geology (Morris, 1976, p. 53).

Among those "competent Biblical scholars" to whom Dr. Morris referred are the following. Wilbur M. Smith, former dean of the Moody Bible Institute, wrote: "First of all, we must dismiss from our mind any conception of a definite period of time, either for creation itself, or for the length of the so-called six creative days" (1945, p. 312). Bernard Ramm labeled the belief that the days of creation were 24-hour periods the "naive, literal view" (1954, pp. 120-121). Merrill Unger suggested that the view which understands the days of Genesis 1 to be literal 24-hour days is "generally recognized as untenable in an age of science" (1966, p. 38). Kenneth Taylor, producer of the *Living Bible Paraphrased*, added footnotes to the text of Genesis 1 in that volume, explaining that the

Hebrew phrase "evening and morning" actually meant a long period of time. In his book, *Evolution and the High School Student*, Taylor wrote:

> To me it appears that God's special creative acts occurred many times during 6 long geological periods capped by the creation of Adam and Eve perhaps more than 1 million years ago. This idea seems to do justice both to the Bible and to what geologists and anthropologists currently believe. If they change their dates up or down, it will make no difference to this belief, unless to move Adam's age forward or backward (1974, p. 62).

Edward John Carnell of Fuller Theological Seminary advised: "And since orthodoxy has given up the literal-day theory out of respect for geology, it would certainly forfeit no principle if it gave up the immediate-creation theory out of respect for paleontology. The two seem to be quite parallel" (1959, p. 95).

In more recent times, the Day-Age Theory has been championed by such writers as Davis A. Young (*Creation and the Flood*, 1977, p. 132), Alan Hayward (*Creation and Evolution: The Facts and the Fallacies*, 1985, p. 164), Howard J. Van Till (*The Fourth Day*, 1986, pp. 75-93; *Portraits of Creation*, 1990, pp. 236-242), and Hugh Ross (*Creation and Time*, 1994, pp. 45-90). Others have lent it their support as well. Jack Wood Sears, mentioned earlier, is on record as advocating this view.[*] In his audio-taped lecture, *Questions and Answers: Number One*, John Clayton remarked:

---

[*] In December 1977, Dr. Sears and I shared the platform at a week-long series of lectures in Zimbabwe (then Rhodesia), Africa. During the lectures in Salisbury, in responding to a question from the audience I stated that the days of creation in Genesis 1 were literal, 24-hour periods. As I returned to my chair, Dr. Sears leaned over to me and said, "There's not a Hebrew scholar in the world who would agree with you on that point. You are very much mistaken in believing the days of creation to be 24-hour days." In his lecture the following day, he publicly took issue with my comment that the days were of a 24-hour duration. In my debate with him in Denton, Texas in November 1983, he once again made clear his position that likely the days of Genesis were long epochs or ages of time.

I believe it is totally inconsequential as to whether or not the days of Genesis were 24-hour days or not. It isn't until the fourth day until the Sun and Moon were established as chronometers. There were no days, seasons, etc.—at least as we know them—before the fourth day! (n.d.[c]).

There are some, however, who are quite cautious not to reveal their own predisposition toward the Day-Age Theory, and who go to great lengths to suggest that this is best left an "open matter" because there are "good arguments on both sides of the issue." Burton Coffman took just such a position in his *Commentary on Genesis* (1985).

There are still others who go through the motions of appearing to be "neutral," when in fact they clearly are not. For example, in the April 4, 1986 edition of *Gospel Minutes* (a weekly publication among the churches of Christ), co-editor Clem Thurman spent a page-and-a-half answering a reader's question on whether or not these days were to be considered as literal days (1986, 35[14]:2-3). He gave three brief points (using approximately two column inches of space) as to why the days "might" be considered as literal, and almost **two full columns** giving reasons why they should not. Then, of course, he urged each reader to "decide for himself" what the "correct" answer might be. Why not just be honest and openly advocate the Day-Age Theory without going through all these machinations?[*]

---

[*] One of the strangest concepts set forth regarding the days of Genesis 1 has been suggested by Gerald L. Schroeder in his book, *Genesis and the Big Bang*. "God might have plunked man down in a world that was ready-made from the instant of creation. But that was not on the Creator's agenda. There was a sequence of events, a development in the world, which led to conditions suitable for man. This is evident from the literal text of Genesis 1:1-31. **By God's time frame, the sequence took six days. By our time frame, it took billions of years**" (1990, p. 85, emp. added; see also Schroeder, 1997). Most readers no doubt will wonder how, by "God's time," it took six days, yet in "our time" it took billions of years? Why is it that writers cannot be forthright and admit that they have no intention of believing the "literal text" (to use Schroeder's own words) of the Genesis account as it is written?

DO ALL THOSE WHO ADVOCATE THE DAY-AGE THEORY BELIEVE IN EVOLUTION?

Is it the case that all those who advocate the Day-Age Theory are either evolutionists or theistic evolutionists? No, not necessarily. There are some who prefer to be called simply "old-Earth creationists" because they claim to accept neither evolution nor theistic evolution. As Surburg observed:

> Many Christians who today hold this view are not necessarily evolutionists. They do not believe that God employed the evolutionary method to produce man, and they endeavor to reconcile the process indicated by paleontology with the creative days of Genesis (1959, p. 57).

Williams agreed, but cautioned:

> ...we do not mean to imply that all who hold to the day-age theory are evolutionists. We do insist, however, that such a view can be maintained only by an acceptance of the mental construct known as the geologic column, which is based upon the assumption of evolution (1970, p. 25).

Indeed, if evolutionary dogma (with its accompanying uniformitarian-based dating methods) had not been allowed to sit in judgment on the biblical record, there would have been no need for the Day-Age Theory in the first place. As Williams went on to point out, there also is an inherent danger in accepting such a theory.

> The day-age theory, though espoused by some men who are sincere Christians, is fraught with dangerous consequences to the Christian faith. This question is not merely academic, as some assert, but it directly affects biblical theology.... The first chapters of Genesis must be regarded as the seed plot of the entire Bible, and if we err here, there is reason to believe that those who come under false interpretations of the Genesis account of creation will sooner or later become involved in error in other areas of divine revelation. It is our conviction that **once the interpretation of the six days of creation which makes them extended periods of perhaps millions of years in duration is accepted, the door is opened for the entire evolutionary philosophy** (1970, pp. 24-25, emp. added).

Henry Morris was correct when he said: "The day-age theory is normally accompanied by either the theory of theistic evolution or the theory of progressive creation. ...neither theistic evolution nor progressive creation is tenable Biblically or theologically. Thus the day-age theory must likewise be rejected" (1974, p. 222). Weston W. Fields said that he has noticed

> ...the underlying presupposition of the day-age theory is that geologic evolutionists are correct in their allegations about the immense eons of time necessary to account for the geological features of the earth, and the biological evolutionists are at least partially correct when they say that in some sense higher forms came from lower forms. Thus, many (though not all) day-age theorists are also theistic evolutionists and progressive creationists (1976, p. 166).

IS THERE LEXICAL/EXEGETICAL EVIDENCE TO SUPPORT THE DAY-AGE THEORY?

In examining whether or not there is lexical and exegetical support for the Day-Age Theory, the question should be asked: "If the author of Genesis wanted to instruct his readers on the fact that all things had been created in six literal days, what words might he have used to convey such a thought?" Henry Morris has suggested:

> ...the writer would have used the actual words in Genesis 1. If he wished to convey the idea of long geological ages, however, he could surely have done it far more clearly and effectively in other words than in those which he selected. It was clearly his intent to teach creation in six literal days.

> Therefore, the only proper way to interpret Genesis 1 is not to "interpret" it at all. That is, we accept the fact that it was meant to say exactly what it says. The "days" are literal days and the events described happened in just the way described (1976, p. 54).

A second question that must be asked is this: "Is there lexical and exegetical evidence to suggest that the days of creation should be interpreted as ages of time?" The most thorough rebuttal of the

Day-Age Theory (and the Gap Theory) ever put into print is Weston W. Fields' book, *Unformed and Unfilled*. In that work, Dr. Fields addressed the lack of evidence—from the biblical text itself —for the Day-Age Theory.

> With the Gap Theory the Day-Age Theory shares the advantage of allowing unlimited amounts of time. But it also has an advantage which the Gap Theory does not: it allows the geologist the **sequence** he wants (assuming he ignores the biblical sequences), and it allows the biologists to have partial or complete evolution. However, it also shares one disadvantage with the Gap Theory—indeed, it outdoes the Gap Theory in this particular: it rests on very scanty exegetical evidence. The lexical exility on which it is based is almost unbelievable; consequently, we must conclude that it springs from presuppositions—a fact transparent even to the casual reader (1976, pp. 156-166, emp. in orig.).

Fields then proceeded to present the lexical evidence.

> ...As in the case of other problems involving meanings of words, our study must begin with Hebrew lexicography. Nearly all the defenders of the theory fail, however, to give any lexical backing to the theory. The reader is left completely uninformed concerning the use of *yom* (day) in the Old Testament. Therefore, we have listed a complete summary of both Brown, Driver, and Briggs's as well as Koehler and Baumgartner's listings. Nothing less than a **complete** examination of the evidence will suffice. In the lexicon of Brown, Driver, and Briggs, there are seven primary meanings for *yom* (day), with numerous subheadings:
>
> 1. Day, opposite of night. Listed under this heading are Genesis 1:5,15,16,18.
>
> 2. Day, as a division of time.
>
>   a. working day.
>
>   b. a day's journey.
>
>   c. to denote various acts or states such as seven days, Genesis 7:4.

d. day as defined by evening and morning. Listed here are Genesis 1:5,8,13,19,23,31.

e. day of the month.

f. day defined by substantive, infinitive, etc., such as the "snowy day."

g. particular days defined by proper name of place, such as the Sabbath Day.

h. your, his, or their day, as in the sense of the day of disaster or death: "your day has come."

3. The day of Yahweh, as the time of coming judgment.

4. The days of someone, equaling his life, or his age: "advanced in days."

5. Days.

a. indefinite: some days, a few days.

b. of a long time: "many days."

c. days of old: former or ancient times.

6. Time.

a. vividly in general sense as in the "time of harvest."

b. used in apposition to other expressions of time, such as a "month of days" equals a "month of time."

7. Used in phrases with and without the prepositions.

a. such as with the definite article, meaning "today."

b. in the expression "and the day came" meaning "when."

c. in an expression such as "lo, days are coming."

d. in construct before verbs, both literally, **the day of**, and (often) in general sense—**the time of** (forcible and pregnant representing the act vividly as that of a single day). Under this definition is listed Genesis 2:4.

e. day by day (yom yom).

f. in expression such as "all the days" meaning always, continually.

g. in an additional phrase with *bet* meaning **on** a particular day.

h. with *kap*, meaning as, like the day.

i. with *lamed*, meaning on or at the day.

j. with *min*, meaning since the day or from the day.

k. with *lemin*, meaning since the day.

l. with *'ad*, meaning until the day.

m. with *'al*, meaning upon the day.

Koehler and Baumgartner list the following usages of *yom*:

1. Day, bright daylight, as opposite of night.

2. Day, of 24 hours. Listed under this heading is Genesis 1:5.

3. Special days, such as the "day of prosperity," or the "day of adversity."

4. Yahweh's day.

5. Plural or day, such as "seven days."

6. Plural of day, such as "the days of the years of your life."

7. Plural of day in a usage to refer to a month or year.

8. Dual, such as in the expression, "a day or two."

9. With the article, "that day."

10. With a preposition such as *bet*, "on the day," or "when."

Now **these** are the meanings the lexicons give. For the reader interested in **all** the evidence, here it is. We must immediately raise the question: where is the lexical support for identifying the days of Genesis as **long periods of time**? Far from supporting the notion that the creative days of Genesis 1 are vast ages, extending, perhaps, over millions of years, the lexicons suggest that "day," as used to refer to creation is of the normal 24 hours duration. This is the **natural** interpretation (1976, pp. 169-172, emp. in orig.).

The evidence supporting the days of creation being 24-hour periods is overwhelming, as Fields has documented. In addition to his evidence, I would like to offer the following for consideration.

EVIDENCE SUBSTANTIATING THE DAYS OF GENESIS 1 AS LITERAL, 24-HOUR PERIODS

1. The days of creation should be accepted as literal, 24-hour periods because the context demands such a rendering.

> The language of the text is simple and clear. Honest exegetes cannot read anything else out of these verses than a day of 24 hours and a week of 7 days. There is not the slightest indication that this is to be regarded as poetry or as an allegory or that it is not to be taken as a historical fact. The language is that of normal human speech to be taken at face value, and the unbiased reader will understand it as it reads. There is no indication that anything but a literal sense is meant (Rehwinkle, 1974, p. 70).

It is true that the word in the Hebrew for day (*yom*), as in other languages, is employed with a variety of meanings. But, as in other languages, context is critical in determining exactly what the word means in any given instance. Morris has noted:

> There is no doubt that *yom* can be used to express time in a general sense. In fact, it is actually translated as "time" in the King James translation 65 times. On the other hand, it is translated as "day" almost 1200 times ....Whenever the writer really intended to convey the idea of a very long duration of time, he normally used some such word as *'olam* (meaning "age" or "long time") or else attached to *yom* an adjective such as *rab* (meaning "long"), so that the two words together *yom rab*, then meant "long time." But *yom* by itself can apparently never be proved, in one single case, to **require** the meaning of a long period of time, and certainly no usage which would suggest a geologic age (1974, p. 223, emp. in orig.).

The following quotation from Arthur Williams documents several important points in this controversy, especially in light of the Day-Age theorists' inconsistency. We are told that *yom* in Genesis 1 is an "age." Yet Day-Age proponents are unwilling to translate the word in this fashion elsewhere, for it makes no sense to do so and destroys the meaning of the passages.

> What did the word *yom* (day) mean to Moses and to Israel in the day in which the books of Moses were written?...
>
> In the Genesis account of creation the word "day" occurs 14 times, always a translation of the Hebrew word *yom*. Those who hold to the day-age theory ask us to give the word "day" a meaning which it has nowhere else in the five books of Moses....
>
> As if the consistent significance of the word *yom* throughout the writings of Moses were not enough to establish the meaning of the English word "day," God added statements which are difficult to interpret otherwise. "...God divided the **light** from the **darkness**. And God called the light **Day** and the darkness he called **Night**. And the **evening** and the **morning** were the first day." In the light of cultural considerations of hermeneutics, can anyone honestly believe that these terms as used in the Genesis account of creation had a meaning almost infinitely removed from the meaning which they had elsewhere in the writings of Moses? The word "day" would have had no meaning to Moses or to his contemporaries other than that which was limited by reference to the sun. It would be impossible to prove from Scripture that the Israelites in the days of Moses had any concept of a "day" in terms of millions or billions of years. The evidence arising from serious consideration of the cultural meaning of the word *yom* as used by Moses and understood by the Israelites is wholly on the side of the 24-hour day in the Genesis account of creation. Such a view is consistent with its meaning as used by Moses throughout his writings (1970, pp. 26-28, emp. in orig.).

As an example of the point Dr. Williams is making, consider the use of *yom* in Numbers 7:12,18. In this context, the discussion is the offering of sacrifices by the princes of Israel. Verse 12 records: "And he that offered his oblation the first day was Nahshon, the son of Amminadab, of the tribe of Judah." Verse 18 records, "On the second day Nethanel the son of Zuar, prince of Issachar, did offer." Notice the sequential nature involved via the use of "first day," and "second day." Do Day-Age theorists suggest that Moses meant to say "in the first eon," or "in the second age" these events transpired? Of course not. Why, then, should the treatment of the word *yom* in Genesis 1 be any different? Indeed, it would not be, were it not for the desire to incorporate evolutionary theory into the biblical text.

2. The days of creation should be accepted as literal, 24-hour periods because God both **used and defined** the word *yom* in the context of Genesis 1. It is nothing short of amazing to discover the evidence built into the text for "interpreting" what kind of days these were. In Genesis 1:5, Moses wrote: "And God called the light Day, and the darkness He called Night. And the evening and the morning were the first day." The "first day" thus is defined as a period of both day and night— i.e., a normal day.

    Further, Genesis 1:14 is instructive in this matter: "And God said, Let there be lights in the firmament of heaven to divide the day from the night; and let them be for signs and for seasons, and for days and for years." If the "days" are "ages," then **what are the years**? If a day is an age, then what is a night? The whole passage becomes ridiculous when one "reinterprets" the word "day." Marcus Dods, writing in the *Expositor's Bible*, said simply: "If the word 'day' in this chapter [Genesis 1—BT] does not mean a period of 24 hours, the interpretation of Scripture is hopeless" (1948, 1:4-5). Klotz correctly observed:

It is a general principle of Biblical interpretation that a word is to be taken in its everyday meaning unless there is compelling evidence that it must be taken in a different sense.... But there is nothing in the text or context of Genesis 1 which indicates that these were long periods of time. Sound principles of Biblical interpretation require that we accept this "day" as being an ordinary day (1955, pp. 84-85).

Fields has summarized the argument by stating: "The farther we read in the creation account, the more obvious it is that Moses intended his readers to understand that God created the universe in six 24-hour days. Nothing could be **more** obvious!" (1976, p. 174, emp. in orig.).

3. The days of creation should be accepted as literal, 24-hour periods because whenever *yom* is preceded by a numeral in Old Testament non-prophetical literature (viz., the same kind of literature found in Genesis 1), it **always** carries the meaning of a normal day. Arthur Williams spoke to this point in the *Creation Research Annual* when he said: "We have failed to find a single example of the use of the word 'day' in the entire Scripture where it means other than a period of twenty-four hours when modified by the use of the numerical adjective" (1965, p. 10). Henry Morris has concurred:

It might still be contended that, even though *yom* never **requires** the meaning of a long age, it might possibly **permit** it. However, the writer of the first chapter of Genesis has very carefully guarded against such a notion, both by modifying the noun by a numerical adjective ("first day," "second day," etc.), and also by indicating the boundaries of the time period in each case as "evening and morning." Either one of these devices would suffice to limit the meaning of *yom* to that of a solar day, and when both are used, there could be no better or surer way possible for the writer to convey the intended meaning of a literal solar day.

To prove this, it is noted that whenever a limiting numeral or ordinal is attached to "day" in the Old Testament (and there are over 200 such instances), the meaning is always that of a literal day (1974, pp. 223-224, emp. and parenthetical comment in orig.).

Raymond Surburg was invited to contribute to the book, *Darwin, Evolution, and Creation*, edited by Paul Zimmerman. In his chapter, Dr. Surburg quoted from a letter written by the renowned Canadian anthropologist, Arthur C. Custance and sent to nine contemporary Hebrew scholars, members of the faculties of nine leading universities—three each in Canada, the United States, and England. In his letter, Dr. Custance inquired about the meaning of *yom* as used in Genesis. For example, he asked: "Do you understand the Hebrew *yom*, as used in Genesis 1, accompanied by a numeral, to be properly translated as: (a) a day as commonly understood, or (b) an age, or (c) an age or a day without preference for either?" Seven of the nine replied, and all stated that the word *yom* means "a day as commonly understood" (as quoted in Surburg, 1959, p. 61). Thus, when the writer states in Exodus 20: 11 that God created the Earth and everything in it in six days, he meant what he said—six literal, 24-hour days.

4. The days of creation should be accepted as literal, 24-hour periods because whenever *yom* occurs in the plural in Old Testament non-prophetical literature (viz., the same kind of literature found in Genesis 1), it always carries the meaning of a normal day. *Yamim* (Hebrew for "days") appears over 700 times in the Old Testament. In each instance where the language is non-prophetical in nature, it **always** refers to literal days. Thus, in Exodus 20:11, when the Scriptures say that "in six days the Lord made heaven and earth, the sea, and all that in them is," there can be no doubt that six literal days are meant. Even the most liberal Bible scholars do not attempt to negate the force of this argument by suggesting that Genesis 1 and Exodus 20:11 are prophetical.

5. The days of creation should be accepted as literal, 24-hour periods because whenever *yom* is modified by the phrase "evening and morning" in Old Testament non-prophetical literature (viz., the same kind of literature found in Genesis 1), it always carries the meaning of a normal day.

> Having separated the day and night, God had completed His first day's work. "The evening and the morning were the first day." This same formula is used at the conclusion of each of the six days; so it is obvious that the duration of each of the days, including the first, was the same.... It is clear that, beginning with the first day and continuing thereafter, there was established a cyclical succession of days and nights—periods of light and periods of darkness.
>
> The writer not only defined the term "day," but emphasized that it was terminated by a literal evening and morning and that it was like every other day in the normal sequence of days. In no way can the term be legitimately applied here to anything corresponding to a geological period or any other such concept (Morris, 1976, pp. 55-56).

Addressing the text from the perspective of a Christian scholar who had studied biblical languages for more than fifty years, Guy N. Woods wrote:

> The "days" of Genesis 1 are divided into light and darkness exactly as is characteristic of the day known to us. "And God saw the light, that it was good; and God divided the light from the darkness. And God called the light Day, and the darkness He called Night. **And the evening and the morning were the first day**" (Genesis 1:4,5). This simple and sublime statement is decisive of the matter. Of what was the first day composed? Evening and morning. Into what was it divided? Light and darkness. The Hebrew text is even more emphatic. The translation, "And the evening and the morning were the first day" is literally, "And evening was, and day was, day one." The two periods—evening and morning—

made one day. The Jewish mode of reckoning the day was from sunset to sunset; i.e., evening and morning, the two periods combining to make **one** day (1976, p. 17, emp. in orig.).

This phrase "evening and morning" is important as a modifier, especially in light of the fact that Day-Age theorists insist that these days were long epochs of time. Has anyone ever seen an "eon" with an evening and morning?

Some have suggested that literal, 24-hour days would have been impossible until at least the fourth day because the Sun had not been created yet. Notice, however, that the same "evening and morning" is employed **before** Genesis 1:14 (the creation of the Sun) as after it. Why should there be three long eras of time before the appearance of the Sun, and only 24-hour days after its creation? Both Klotz and Woods have addressed this objection.

> Insofar as the view is concerned that these could not be ordinary days because the sun had not been created, we should like to point to the fact that we still measure time in terms of days even though the sun does not appear or is not visible. For instance, north of the Arctic Circle and south of the Antarctic Circle the sun does not appear for periods of time up to six months at the poles themselves. We would not think of measuring time in terms of the appearance or lack of appearance of the sun in these areas. No one would contend that at the North or South Pole a day is the equivalent of six months elsewhere (Klotz, 1955, p. 85).

> ...If to this the objection is offered that the sun did not shine on the earth until the fourth day, it should be remembered that it is the function of the heavenly bodies to **mark** the days, not **make** them! It is night when no moon appears; and the day is the same whether the sun is seen or not (Woods, 1976, p. 17, emp. in orig.).

By way of summary, it may be said that:

(a) The phrase "evening and morning" was the Hebrew way of describing a literal, 24-hour day.

(b) There are no instances in the non-prophetical Old Testament passages where the phrase "evening and morning" represents anything more than a literal, 24-hour day.

(c) The presence of the Sun and Moon do not regulate the day and the night. The Earth's rotation on its axis does that. Since the phrase "evening and morning" is used both before and after the Sun's creation, the days are obviously literal, 24-hour days.

6. The days of creation should be accepted as literal, 24-hour periods because Moses had at his disposal the means by which to express long periods of time, yet purposely did not use wording in the original Hebrew which would have portrayed that idea. Fields has commented:

> Perhaps the most telling argument against the Day-Age Theory is, "what else could God say to convey the idea that the days of creation were **literal** days?" He used the **only** terms available to him to communicate that idea. There was a word, on the other hand, which Moses could have used had he wanted to signify **ages** or vast **periods of time**. He could have used the word *dor* which has that very meaning. But instead he used the word "day," and we think the reason he did is very obvious to the unbiased reader: He wanted to tell his readers that all of creation took place in six literal 24-hour days! (1976, pp. 177-178, emp. in orig.).

7. The days of creation should be accepted as literal, 24-hour periods because of the problems in the field of botany if the days are stretched into long periods of time. Woods wrote:

> Botany, the field of plant-life, came into existence on the third day. Those who allege that the days of Genesis 1 may have been long geological ages, must accept the absurd hypothesis that plant-life survived in periods of total darkness through half of each geologic age, running into millions of years (1976, p. 17).

Henry Morris also has addressed this issue:

> The objection is sometimes raised that the first three days were not days as they are today since the sun was not created until day four. One could of course turn this objection against those who raise it. The longer the first three days, the more catastrophic it would be for the sun not to be on hand during those days, if indeed the sun is the only possible source of light for the earth. The vegetation created on the third day might endure for a few hours without sunlight, but hardly for a geologic age! (1974, p. 224).

In addition, there is a serious problem regarding reproduction of plants. The Genesis text indicates that plants were created on day three. Yet other living things were not created until days five and six. How could plants have survived that are pollinated solely by insects? Clover is pollinated by bees, and the yucca plant has the pronuba moth as its only means of pollination. How did plants multiply if they were growing millions of years before the insects came into existence?

8. The days of creation should be accepted as literal, 24-hour periods because of plain statements about them within the Scriptures.

    (a) "for in six days Jehovah made heaven and earth, the sea, and all that in them is" (Exodus 20:11).

    (b) "For He spake, and it was done; He commanded, and it stood fast" (Psalm 33:9).

    (c) "Let them praise the name of Jehovah; for he commanded and they were created" (Psalm 148:5).

    (d) "for in six days Jehovah made heaven and earth, and on the seventh day he rested, and was refreshed" (Exodus 31:17).

Does a simple, straightforward reading of these verses imply a long period of evolutionary progress, or six literal, 24-hour days and instantaneous creation? Riegle has written:

The Hebrew text implies that the Creative acts were accomplished instantly. In Genesis 1:11 God's literal command was, "Earth, sprout sprouts!" In the very next verse we find the response to the command—"The earth caused plants to go out." There is no hint that great ages of time were required to accomplish this phase of the Creation. It could have been done in only minutes, or even seconds, as far as God's creative power is concerned (1962, pp. 27-28).

In its appropriate context, each of these passages can be understood correctly to be speaking only of literal days and instantaneous creation.

9. The days of creation should be accepted as literal, 24-hour periods because of God's explicit command to the Israelites to work six days and rest on the seventh, just as He had done. An important fact that should not be overlooked in this particular context is that God not only told His people **what** to do (rest on the seventh day), but **why** to do it (because that is exactly what He did during His week of creative acts).

> Remember the sabbath day, to keep it holy. Six days shalt thou labor, and do all thy work; but the seventh day is a sabbath unto Jehovah thy God: in it thou shalt not do any work, thou, nor thy son, nor thy daughter, thy manservant, nor thy maidservant, nor thy cattle, nor thy stranger that is within thy gates: for in six days Jehovah made heaven and earth, the sea, and all that in them is, and rested the seventh day: wherefore Jehovah blessed the sabbath day, and hallowed it (Exodus 20:8-11).

The Sabbath command can be understood properly only when the days of the week are considered as 24-hour days. Wilder-Smith has summarized the difficulty, in regard to the Sabbath, if the days are not literal, 24-hour days.

> Another difficulty arises if one tries to apply the age-equals-day interpretation. The whole important biblical doctrine of the Sabbath is weakened by this view. For

God is reported as having rested on the seventh day after working the six days. The implication is that man should also rest on the seventh day as God did. But did God rest for an age, maybe of millions of years? The whole biblical concept of the Sabbath is coupled with six working days and one day of rest in seven. God certainly did not need to rest, but presumably set us a pattern with the Sabbath rest (1975, p. 44).

OBJECTIONS CONSIDERED

Three specific objections to 24-hour creation days often are mentioned by those who advocate an old Earth.

"One Day Is With The Lord As A Thousand Years"

The first objection has to do with the passage found in 2 Peter 3:8: "But forget not this one thing, beloved, that one day is with the Lord as a thousand years, and a thousand years as one day." This passage is used by proponents of the Day-Age Theory to suggest that the days of Genesis 1 could have been long ages or epochs of time since, according to Peter, one day is "as a thousand years." Woods has responded:

The passage should be considered in the light of its context. The material heavens and earth are to suffer destruction by fire, despite the mockers who scoff at such predictions and who allege, in the face of the earth's earlier destruction by water, that all things must continue as they are from the beginning (2 Peter 3:1-7). All such are "willingly ignorant," and refuse to accept the clear and obvious lessons of history. Faithful followers of the Lord are not to be influenced by these skeptics, but to remember "that one day is with the Lord as a thousand years, and a thousand years as one day."

By this the apostle meant that the passing of time does not, in any way, effect the performance of God's promises or threats. He is not influenced by the passing of the centuries; and the lapse of time between the promise or threat, and the performance, is no factor, at all. With man, it definitely is. That which we promise to do tomorrow, we are much more likely to do, than that which we promise next year, or in the next century, since we may not be here then to fulfill the promise. But, this limitation, so characteristic of man, does not influence Deity. The passing of a thousand years, to God, does not alter his plans and purposes any more than a day, and he will carry them out as he has planned, regardless of the amount of time which is involved (1976, p. 146).

The discussion in 2 Peter 3:8 simply means that time is of little essence with God. Peter's obvious intent is to teach that God does not tire, though thousands of years may pass, because with Him a thousand years are **as** a day. [Notice that the text does not say a day **is** a thousand years; rather, it says a day is **as** a thousand years.] This passage serves to illustrate the eternal nature of God and His faithfulness to His promises—not that the days of Genesis 1 are "eons of time." The days of Genesis 1 are not to be reinterpreted by misapplying the message of 2 Peter 3:8.

Too Much Activity On Day Six

The second objection to the days of Genesis being literal, 24-hour periods is that the sixth day could not have been a normal day because too much activity occurred on that day. Alan Hayward, who accepts this criticism as legitimate because he holds to the Day-Age Theory, has explained why he believes this to be a valid argument against 24-hour days.

Finally, there is strong evidence that the sixth day of creation must have lasted more than 24 hours. Look how much took place in that sixth day! To begin with, God created the higher animals, and then created Adam. After that:

And the Lord God planted a garden in Eden.... And out of the ground the Lord God made to grow every tree... (Genesis 2:8,9). Then **every** living animal and **every** bird was brought to Adam for naming.

In all that long procession of living things, Adam saw that "there was not found a helper fit for him" Genesis 2:20). So God put Adam to sleep, created Eve, and presented her to Adam, who joyfully declared:

This **at last** is bone of my bones and flesh of my flesh; she shall be called Woman, because she was taken out of Man (Verse 23).

All commentators are agreed that the expression translated "at last" in the RSV means just that. They usually express the literal meaning of the Hebrew as "now, at length," and some of them quote numerous other passages in the Old Testament where this Hebrew word carried the same sort of meaning. Thus, the Hebrew indicates that Adam had been kept waiting a long time for his wife to appear—and all on the sixth day (1985, pp. 164-165, emp. in orig.).

This is one of the few attempts to prove that the days of creation were long periods of time by actually appealing to the Bible itself. Generally no such attempts are made by those holding to the Day-Age Theory. Instead, they routinely base their case on scientific arguments, appealing to the apparent antiquity of the Earth, geological phenomena, etc. Here, however, their position is as follows: (1) there is textual evidence in Genesis 2 that the sixth day of creation could not have been a literal day (as suggested by Hayward, above); (2) but obviously it was the same type of "day" as each of the previous five; (3) thus, none of the "days" of the creation week is to be viewed as literal.

The argument (from Hayward's statement of it) is two-pronged. First, it is said that after God created Adam on the sixth day, He commissioned him to name the animals **before Eve was fashioned** later on that same day—which would have taken a much longer period than a mere 24-hour day. Second, it is alleged that when Adam first saw Eve, he said, "This is **now** [Hayward's "at last"] bone of my bones...," and his statement thus reflects that he had been some time without a mate—certainly longer than a few hours. This compromise is advocated not only by Hayward, but by Gleason Archer in his *Encyclopedia of Biblical Difficulties* (1982, pp. 58ff.) and by Hugh Ross in *Creation and Time* (1994, pp. 50-51).

Significantly, professor Archer reveals that he has been influenced by the assertions of evolutionary geochronology. His discussion of this matter is in response to the question: "How can Genesis 1 be reconciled with the immense periods of time indicated by fossil strata?" He has claimed that there is conflict between Genesis and the beliefs of evolutionary geologists only if one understands "Genesis 1 in a completely literal fashion," which, he asserts, is unnecessary. Dr. Archer has suggested that "God gave Adam a major assignment in natural history. He was to classify every species of animal and bird found in the preserve" (1982, p. 59). He further stated that it

> ...must have taken a good deal of study for Adam to examine each specimen and decide on an appropriate name for it, especially in view of the fact that he had absolutely no human tradition behind him, so far as nomenclature was concerned. It must have required some years, or, at the very least, a considerable number of months for him to complete this comprehensive inventory of all the birds, beasts, and insects that populated the Garden of Eden (1982, p. 60).

One would be hard pressed to find a better example of "the theory becoming father to the exegesis" than this.[*] Archer has simply "read into" the divine narrative the assumptions of his baseless view. Let us take a careful look at the Bible **facts**.

First, apparently only those animals that God "brought" unto Adam were involved, and this seems to be limited, as Archer concedes, to Eden. Second, certain creatures were excluded. There

---

[*] I wish to thank Wayne Jackson for permission to edit and reproduce portions of this material from *Reason & Revelation*, the monthly journal on Christian evidences that he and I formerly co-edited (and for which I currently serve as editor).

is, for example, no mention of fish or creeping things. Third, the text does not suggest how broad the categories were that Adam was to name. It is sheer assertion to claim that he was to name all "species." God created living organisms according to "kinds," which, in the Bible appears to be a rather elastic term. It translates the Hebrew word *min*, which sometimes seems to indicate species, sometimes genus, and sometimes family or order. [But, as Walter C. Kaiser, Chairman of the Department of Old Testament and Semitic Languages, Trinity Divinity School, has observed: "This gives no support to the classical evolutionist view which requires developments across kingdom, phyla, and classes" (*see* Harris, et al., 1980, 1:504).] Fourth, why should it be assumed that Adam had to "give a good deal of study" to this situation? He never had to "study" such things as walking, talking, or how to till the ground; clearly Adam had been endowed miraculously with a mature knowledge that enabled him to make his way in that antique environment. He needed no "human tradition" behind him; he was "of God" (*see* Luke 3:38).

Let us examine what some other scholars have said about this. C.F. Keil observed that although Adam and Eve were created on the same day "there is no difficulty in this, since it would not have required much time to bring the animals to Adam to see what he would call them, as the animals of paradise are all we have to think of" (1971, 1:87). H.C. Leupold noted:

> That there is a limitation of the number of creatures brought before man is made apparent by two things. In the first place, the beasts are described as beasts **of the field** (*hassadheh*), not beasts **of the earth**, as in 1:24. Though there is difficulty in determining the exact limits of the term "field" in this instance, there is great likelihood (cf. also v. 5) that it may refer to the garden only. In the second place, the fish of the sea are left out, also in v. 20, as being less near to man. To this we are inclined to add a third consideration, the fact, namely, that the garden could hardly have been a garden at all if all creatures could have overrun it unimpeded.

Since then, very likely, only a limited number of creatures are named, the other difficulty falls away, namely, that man could hardly have named all creatures in the course of a day (1942, pp. 130-131, emp. in orig.).

As Henry Morris has pointed out,

...the created kinds undoubtedly represented broader categories than our modern species or genera, quite possibly approximating in most cases the taxonomic family. Just how many kinds were actually there to be named is unknown, of course, but it could hardly have been as many as a thousand. Although even this number would seem formidable to us today, it should be remembered that Adam was newly created, with mental activity and physical vigor corresponding to an unfallen state. He certainly could have done the job in a day and, at the very most, it would have taken a few days even for a modern-day person, so there is nothing anywhere in the account to suggest that the sixth day was anything like a geological age (1984, p. 129, emp. in orig.).

As it turns out, Dr. Archer's argument about the animals is much ado about nothing.

Archer further suggested that this extended period of naming the animals left Adam with a "long and unsatisfying experience as a lonely bachelor" and so finally he was "emotionally prepared" when Eve arrived. Another writer declared: "It seems that he [Adam-BT] had been searching diligently for a long time for a suitable mate, and when he found her, he burst out, **This at last** [literally, 'this time'] **is bone of my bones**, etc." (Willis, 1979, p. 113, emp. in orig.).

Again, one can only express amazement at how some scholars so adroitly "read between the lines." There is nothing in the statement, "This is now bone of my bones..." that demands a long, lonely bachelorhood for Adam. The Hebrew word translated "now" is pa'am. The term does not require a protracted span of time, as asserted by Willis. It can denote simply a contrast with that which has been recorded previously, as it does in this context. Professor M.W. Jacobus observed that the term denoted "**this time —in this instance**, referring to the other **pairs**," and so simply expressed Adam's satisfaction with his mate in contrast to the animals he had been naming (1864, p. 110, emp. in orig.). Robert Jamieson wrote:

> ...**this time**, is emphatic (cf. 30:30; 46:30). It signifies "now indeed," "now at last," as if his memory had been rapidly recalling the successive disappointments he had met with in not finding, amidst all the living creatures presented to him, any one capable of being a suitable companion to him (1945, 1:46, emp. in orig.).

There is, therefore, nothing in Genesis 2 that is in conflict with the plain, historical statements of Genesis 1:27ff.: "And God created man in his own image, in the image of God created he him; male and female created he them.... And there was evening and there was morning, the sixth day." As I have pointed out repeatedly, the Scriptures indicate that the creation week of six days was composed of the same kind of "days" that the Hebrews employed in their observance of the Sabbath (Exodus 20:8-11), and though this argument has been ridiculed, **it never has been answered**.

There is another point, from the New Testament, that is worthy of consideration. In 1 Timothy 2:13, Paul wrote: "For Adam was first formed, **then** Eve." Of special interest here is the word "then" [Greek, *eita*]. This term is an adverb of time meaning "then; next; after that" (Thayer, 1962, p. 188). It is found 16 times in the New Testament in this sense. [Once it is employed in argumentation to add a new reason and so is rendered "furthermore" (Hebrews 12:9).] The word, therefore, generally is used to suggest a logical sequence between two occurrences and **there is never an indication that a long lapse of time separates the two**. Note the following:

(a) Jesus "girded himself. Then [eita] he poureth water into the basin" (John 13:5).

(b) From the cross, Jesus said to Mary, "Woman, behold thy son! Then [eita] saith he to the disciple..." (John 19:26-27). Compare also John 20:27—"Then [eita] saith he to Thomas..." and Mark 8:25.

(c) In Luke 8:12, some seed fell by the wayside, "then [eita] cometh the devil and taketh away the word from their heart." And note Mark's parallel, "straightway cometh Satan, and taketh away the word..." (4:15). These examples reveal no long lapses of time.

(d) James says a man "is tempted when he is drawn away by his own lust, and enticed. Then [eita] the lust, when it hath conceived, beareth sin" (1:14-15). How long does that take?

(e) Christ appeared to Cephas; "then [eita] to the twelve" (1 Corinthians 15:5) and this was on the same day (Luke 24:34-36). See also 1 Corinthians 15:7.

(f) In speaking of Christ's coming, Paul declares, "Then [eita] cometh the end" (1 Corinthians 15:23-24). Will there be a long span of time (1,000 years), as the millennialists allege, between Christ's coming and the end? Indeed not.

(g) For the other uses of *eita*, see Mark 4:17, Mark 4:28, 1 Corinthians 12:28, and 1 Timothy 3:10.

So, "Adam was first formed, then [eita] Eve" (1 Timothy 2:13). Paul's use of this adverb, as compared with similar New Testament usages elsewhere, is perfectly consistent with the affirmation of Moses that Adam and Eve were made on the same literal day of history.

God's Sabbath Rest Still Is Continuing

Day-Age theorists sometimes suggest that the seventh "day" still is continuing. Their argument is that since "evening and morning" is not mentioned in regard to the seventh day, it must not have been a 24-hour day. Therefore, we are living in the seventh day —a position they must defend to remain consistent. There are, however, a number of serious problems with this approach. The first has been explained by Woods.

> Jehovah finished his labors at the end of the sixth day, and on the seventh rested. The narrative provides no basis for the assumption that the day he rested differed in any fashion from those which preceded it. It evidently was marked out and its length determined in the same manner as the others. If it was not a day of twenty-four hours, it sustains no resemblance to the sabbath which was given to the Israelites (1976, pp. 17-18).

Moses' obvious intent was for the reader to understand that God: (1) rested (past tense); and (2) gave the seventh day (the Sabbath) as a day of rest because He had rested on that day.

There is a second problem with the view that the seventh day still is continuing. James Pilgrim has addressed that problem.

> ...if the "day-age" theorists accept day seven as an "age" also, we ask, "What about day eight, or day nine, or day ten...?" On the **assumption** that the earth is 7,000 years old (a most distinct possibility), let the "day-age" proclaimers put 2,555,000 days (7,000 years at 365 days per year) on a page. Now let them circle the day which began the normal 24-hour day. Let them also give just one scripture reference to substantiate the validity of that circle. Can they do it? No! Will they do it? No! (1976, p. 522, emp. in orig.).

The third problem with the idea that the seventh day is continuing has to do with Adam, as Woods has noted:

> Adam, the first man, was created in the **sixth** day, lived through the **seventh** day, and into at least a portion of the **eighth** day. If these days were long geologic periods of **millions** of years in length, we have the interesting situation of Adam having lived in a portion of **one** age, through the whole of **another** age, and into at

least a portion of a third age, in which case he was many millions of years old when he finally died! Such a view of course is absurd; and so are the premises which would necessitate it (1976, p 18, emp. in orig.).

Whitcomb has done an excellent job of explaining **why** these things are true:

...Genesis 2:2 adds that He rested on the seventh day. That day also must have been literal, because otherwise the seventh day which God blessed and sanctified would have been cursed when God cursed the world and cast Adam and Eve out of the Garden. You see, the seventh day must have ended and the next week commenced before that Adamic curse could have come. Adam and Eve lived through the entire seventh day and into the following week, which is simply a confirmation of the fact that each of the days, including the seventh, was literal (1973a, 2:64-65).

It also has been suggested that Hebrews 4:4-11, where the writer speaks of the **continuation** of God's Sabbath rest, provides support for the Day-Age Theory. First, I would like to present the passage in question along with the argument made from it. Then I would like to offer an explanation of why the passage does not lend credence to the Day-Age Theory and why the argument based on it is faulty. Here is the passage.

For he hath said somewhere of the seventh day on this wise, "And God rested on the seventh day from all his works"; and in this place again, "They shall not enter into my rest." Seeing therefore it remaineth that some should enter thereinto, and they to whom the good tidings were before preached failed to enter because of disobedience, he again defineth a certain day, "Today," saying in David so long a time afterward (even as hath been said before), "Today if ye shall hear his voice, Harden not your hearts." For if Joshua had given them rest, he would not have spoken afterward of another day. There remaineth therefore a sabbath rest for the people of God. For he that is entered into his rest hath himself also rested from his works, as God did from his. Let us therefore give diligence to enter into that rest, that no man fall after the same example of disobedience.

Here is the argument. Proponents of the Day-Age Theory suggest that since God's Sabbath Day (the seventh day of the creation week) continues to this very day, then it follows logically that the other days of the creation week were long periods of time as well (see Ross, 1994, pp. 48-49,59-60; Geisler and Brooks, 1990, p. 230). In support of this position, Hugh Ross wrote: "Further information about the seventh day is given in Hebrews 4. ...we learn that God's day of rest continues" (1994, p. 49).

Wrong! Here is the actual meaning of the passage. While the text speaks clearly of the cessation—beginning on the seventh day —of God's creative activity, the text nowhere suggests that God's seventh day continues from the past into the present. Nor does the passage speak of the duration of the seventh day. Van Bebber and Taylor have addressed this point.

> Like David in the Psalms, the writer of Hebrews is warning the elect not to be disobedient and hard-hearted. Thus, he alludes to Israel in the wilderness who because of their hard hearts could not receive God's promise of rest in Canaan.

> "Rest," as used in these verses by both David and the writer of Hebrews, had a specific historic reference to the promised land of Canaan. The Hebrew word used by David for "rest" was *menuwchah*, which is a general term for rest which has a special **locational** emphasis (e.g., "**the resting place or abode of resting**") [see Brown, et al., 1979, p. 629b]. This concept is echoed by the author of Hebrews who uses the Greek word *katapausis*, which also may refer to an abode or location of resting (Hebrews 4:1,3-5,8).

> At the climax of this passage, the author promises a future day of rest (Hebrews 4:9, Greek: *Sabbatismos*). This is the only time in the New Testament that this word for "rest" is employed. It seems to be a deliberate reference to the Day Seven of Creation. The author does not say, however, that

the seventh day continues on into the future. He uses *Sabbatismos* without an article (like saying **a** Sabbath, rather than **the** Sabbath). In Greek, this grammatical structure would generally represent the character or nature of Day Seven, **without really being Day Seven**. That is, the context makes it clear that the future day of rest will be similar to the original seventh day. The task will be complete; we will live with Christ eternally—our work on earth will be done (1996, pp. 72-73, emp., parenthetical, and bracketed items in orig.).

The passage in Hebrews is using the **essence** of the seventh day of creation to refer to the coming **essence** of heaven—i.e., a place of rest. It is not speaking about the actual **length** of that seventh day.

Furthermore, the fact that God has not been involved in creative activity since the close of day six says absolutely nothing about the duration of the individual days of creation. When God completed the creation, He "rested"—but **only from His work of creation**. He is very much at work now—but **in His work of redemption**, not creation. Jesus Himself said, "My Father worketh even until now" (John 5:17). While it **is correct** to say that God's rest from creative activity continues to this very hour, it **is not correct** to say that His Sabbath Day continues. That was not the Hebrew writer's point, and to suggest that it was represents either a misunderstanding or misuse (or both) of the passage.

God was not saying, via the Hebrew writer, that He wanted to share a **literal** Sabbath Day's rest with His creation. Rather, He was saying that He intended to enjoy a rest that was **typified** by the Sabbath Day's rest. The Israelites who rebelled against God in the wilderness were not able to share either a "rest" by entering into the physical presence of the promised land or a "rest" by entering into the eternal presence of God. Lenski commented on the text as follows:

The point lies in taking all these passages together. The rest from which the Jews of the Exodus were excluded into which we are entering is God's rest, the great Sabbath since the seventh day, of course not of the earthly days and years that have rolled by since then and are still continuing but the timeless, heavenly state that has been established and intended for men in their glorious union with God.

These are not different kinds of rest: the rest of God since creation and a future rest for his people; or a rest into which men have already entered and one that has been established since the redemptive work of Jesus, into which they are yet to enter; or a rest "at the conclusion of the history of mankind." **The seventh day after the six days of creation was a day of twenty-four hours**. On this day God did not create. Thus God made the first seven-day week (Exod. 20:8-11; 31:12-17), and the Sabbath of rest was "a sign" (v. 17) so that at every recurrence of this seventh day Israel might note the significance of this sign, this seventh day of rest being a **type** and a **promise** of the rest instituted for man since the days of creation. Like Canaan, the Sabbath was a type and a promise of this rest (1966, pp. 132-133, emp. added).

Additionally, even if it could be proved somehow that the seventh day of creation were longer than the others (which it cannot), that still would establish only one thing—that the **seventh day** was longer. It would say absolutely nothing about the length of the other six days. And concerning those days, the Bible could not be any clearer than it is in explaining their duration of approximately twenty-four hours. Genesis 1 defines them as periods of "evening and morning" (1:5,8,13,19,23,31). While God's **activity** within each literal day may have been miraculous, there is nothing miraculous about the **length** of the days themselves. They were, quite simply, the same kinds of "days" that we today enjoy. Attempts to reinterpret the message of Hebrews 4 do not alter that fact.

I would like to offer those who are enamored with the Day-Age Theory the following challenge (as set forth by Fields) for serious and thoughtful consideration:

> It is our conclusion, therefore, that the Day-Age Theory is impossible. It is grammatically and exegetically preposterous. Its only reason for existence is its allowance for the **time** needed by the evolutionary geology and biology. We would like to suggest two courses of action for those who so willingly wed themselves to such extravagant misinterpretations of the Scripture: either (1) admit that the Bible and contemporary uniformitarian geology are at odds, reject biblical creation, and defend geological and biological evolution over billions of years; or (2) admit that the Bible and contemporary uniformitarian geology are at odds, study all the geological indications of the **recent creation** of the earth, accept the implications of Noah's flood, and believe the recent creationism of the Bible. One must choose either the chronological scheme of uniformitarianism or the chronological scheme of the Bible, **but the inconsistencies of this sort of interpretation of the Hebrew text for the purpose of harmonizing mutually exclusive and hopelessly contradictory positions can no longer be tolerated** (1976, pp. 178-179, emp. in orig. except for last sentence).

3

THE GAP THEORY

In recent years, the Day-Age Theory has fallen on hard times. Numerous expositors have outlined its shortcomings and have shown that it is without lexical or exegetical support. It has utterly failed to secure the goals and objectives of its advocates—i.e., the injection of geological time into the Genesis account in a biblically and scientifically logical manner, with the subsequent guarantee of an ancient Earth. Therefore, even though it retains its popularity in certain circles, it has been rejected by many old-Earth creationists, theistic evolutionists, and progressive creationists.

Yet the Bible believer who still desires to accommodate his theology to the geologic ages, and to retain his belief in an old Earth, must fit vast time spans into the creation account of Genesis 1 in some fashion. As I explained earlier, there are only three options. The time needed to ensure an old Earth might be placed: (a) **during** the creation week; (b) **before** the creation week; or (c) **after** the creation week. I have shown, in my review of the Day-Age Theory, that the geologic ages cannot be placed into the biblical text **during** the creation week. I now would like to examine the suggestion that they may be inserted **before** the creation week.

For over 150 years, Bible-believers who were determined to insert the geologic ages into the biblical record, yet who realized the inadequacy of the Day-Age Theory to accomplish that task, have suggested that it is possible to place the geologic ages **before** the creation week using what is commonly known as the Gap Theory (also known by such synonyms as the Ruin-and-Reconstruction Theory, the Ruination/Re-creation Theory, the Pre-Adamic Cataclysm Theory, and the Restitution Theory).

Modern popularity of the Gap Theory generally is attributed to the writings of Thomas Chalmers, a nineteenth-century Scottish theologian. Ian Taylor has provided this summary:

> An earlier attempt to reconcile geology and Scripture had been put forward by another Scotsman, Thomas Chalmers, an evangelical professor of divinity at Edinburgh University. He founded the Free Church of Scotland, and because of his outreach to the poor and destitute he later became known as the "father of modern sociology." Traceable back to the rather obscure writings of the Dutchman Episcopius (1583-1643), Chalmers formed an idea, which became very popular and is first recorded in one of his lectures of 1814: "The detailed history of Creation in the first chapter of Genesis begins at the middle of the second verse."[*]
> Chalmers went on to explain that the first statement, "In the beginning God created the Heavens and the Earth and the Earth was without form and void and darkness was on the face of the deep," referred to a pre-Adamic age, about which Scripture was essentially silent. Some great catastrophe had taken place, which left the earth "without form and void" or ruined, in which state it remained for as many years as the geologist required. Finally, approximately six thousand years ago, the Genesis account continues, "The spirit of God moved upon the face of the waters." The re-

---

[*] See: Chalmers, Thomas (1857), "Natural Theology," *The Select Works of Thomas Chalmers*, ed. William Alanna (Edinburgh, Scotland: Thomas Constable), volume five of the twelve volume set.

maining verses were then said to be the account of how this present age was restored and all living forms, including man, created (1984, pp. 362-363).

Through the years, the Gap Theory has undergone an "evolution" of its own and therefore is not easy to define. There are several variations of the Gap Theory, and at times its defenders do not agree among themselves on strict interpretations. I will define the theory as many of its advocates have, recognizing that no single definition can be all-inclusive or encompass all possible facets of the theory. A brief summation of the main tenets of the Gap Theory might be as follows.

The widely held view among Gap theorists today is that the original creation of the world by God, as recorded in Genesis 1:1, took place billions of years ago. The creation then was despoiled because of Satan's disobedience, resulting in his being cast from heaven with his followers. A cataclysm* occurred at the time of Satan's rebellion, and is said by proponents of the Gap Theory to have left the Earth in darkness ("waste and void") as a divine judgment because of the sin of Satan in rebelling against God. The world as God had created it, with all its inhabitants,** was destroyed and left "waste and void," which, it is claimed, accounts for the myriad fossils present in the Earth. Then, God "re-created" (or "restored") the Earth in six literal, 24-hour days. Genesis 1, therefore, is the story of an original, perfect creation, a judgment and ruination, and a re-creation. While there are other minor details that could be included, this represents the essence of the Gap Theory.

---

* It is alleged by some Gap theorists that the cataclysm occurring at Satan's rebellion **terminated** the geologic ages, after which God "re-created" (Genesis 1:2). It is alleged by others that the cataclysm occurred, and **was followed by** the billions of years that constituted the geologic ages; then, at some time determined in the mind of God, He "re-created." Because it is difficult to know which school of thought to follow, both are presented for the reader's consideration.

** Many holding to this theory place the fossils of dinosaurs, so-called "ape-men," and other extinct forms of life in this gap, thereby hoping to avoid having to explain them in the context of God's present creation.

This compromise is popular with those who wish to find a place in Genesis 1 for the geologic ages, but who, for whatever reasons, reject the Day-Age Theory. The Gap Theory is intended to harmonize Genesis and geology on the grounds of allowing vast periods of time between Genesis 1:1 and Genesis 1:2, in order to account for the geologic ages. George H. Pember, one of the earliest defenders of the Gap Theory, wrote:

> Hence we see that geological attacks upon the Scriptures are altogether wide of the mark, are a mere beating of the air. There is room for any length of time between the first and second verses of the Bible. And again; since we have no inspired account of the geological formations we are at liberty to believe they were developed in the order we find them (1876, p. 28).

The *Scofield Reference Bible,*[*] in its footnote on Genesis 1:11, suggested: "Relegate fossils to the primitive creation, and no conflict of science with the Genesis cosmogony remains" (1917, p. 4). Harry Rimmer, in *Modern Science and the Genesis Record* (1937), helped popularize the Gap Theory. Anthropologist Arthur C. Custance produced *Without Form and Void* (1970)—the text that many consider the ablest defense of the Gap Theory ever put into print. George Klingman, in *God Is* (1929), opted for the Gap Theory, as did Robert Milligan in *The Scheme of Redemption* (1972 reprint). George DeHoff advocated the Gap Theory in *Why We Believe the Bible* (1944), and J.D. Thomas, stated in his text, *Evolution and Antiquity,* that "no man can prove that it is not true, at least in part (1961, p. 54). John Clayton has accepted almost all of the Gap Theory, but has altered it to suit his own geological/theological purposes. The end result is an extremely unusual hybrid known as the Modified Gap The-

---

* First published in 1909, by 1917 the *Scofield Reference Bible* had placed the Gap Theory into the footnotes accompanying Genesis 1; in more recent editions, references to the theory may be found as a footnote to Isaiah 45.

ory* (see: Clayton, 1976a, pp. 147- 148; Thompson, 1977, pp. 192-194; 1995, pp. 193-206; McIver, 1988, 8[3]:1-23; Jackson and Thompson, 1992, pp. 114-130).

SUMMARY OF THE GAP THEORY

Those who advocate the Gap Theory base their views on several arguments, a summary of which is given here; comments and refutation will follow.

1. Gap theorists suggest that two Hebrew words in the creation account mean entirely different things. Gap theorists hold to the belief that *bara* (used in Genesis 1:1,21,27) means "to create" (i.e., *ex nihilo* creation). *Asah*, however, does not mean "to create," but instead means "to re-create" or "to make over." Therefore, we are told, the original creation was "created"; the creation of the six days was "made" (i.e., "made over").

2. Gap theorists suggest that the Hebrew verb *hayetha* (translated "was" in Genesis 1:2) should be rendered "became" or "had become"—a translation required in order to suggest a change of state from the original perfect creation to the chaotic conditions implied in verse 2.

3. Gap theorists believe that the "without form and void" of Genesis 1:2 (Hebrew *tohu wabohu*) can refer only to something once in a state of repair, but now ruined. Pember accepted these words as expressing "an outpouring of the wrath of God." Gap theorists believe the cataclysm that occurred was on the Earth, and was the direct result of Satan's rebellion against God. The cataclysm, of course, is absolutely essential to the Gap Theory. Isaiah 14:12-15 and Ezekiel 28:11-17 are used as proof-texts to bolster the theory.

---

* Clayton's Modified Gap Theory will be treated more fully in chapter 4.

4. Gap theorists believe that Isaiah 45:18 ("God created the earth not in vain"—Hebrew, *tohu*; same word as "without form" in Genesis 1:2) is a proof-text that God did **not** create the Earth *tohu*. Therefore, they suggest, Genesis 1:2 can refer only to a judgment brought upon the early Earth by God.

5. Gap theorists generally believe that there was a pre-Adamic creation of both non-human and human forms. Allegedly, Jeremiah 4:23-26 is the proof-text that requires such a position, which accounts for the fossils present in the Earth's strata.

THE GAP THEORY—A REFUTATION

The above points adequately summarize the positions of those who advocate the Gap Theory. I now would like to suggest the following reasons why the Gap Theory should be rejected as false.

1. The Gap Theory is false because of the "mental gymnastics" necessary to force its strained argumentation to agree with the biblical text. Even Bernard Ramm, who championed the idea of progressive creationism, found those mental gymnastics a serious argument against the theory's unorthodox nature.

> It gives one of the grandest passages in the Bible a most peculiar interpretation. From the earliest Bible interpretation this passage has been interpreted by Jews, Catholics, and Protestants as the **original creation of the universe**. In six majestic days the universe and all of life is brought into being. But according to Rimmer's view the great first chapter of Genesis, save for the first verse, is not about original creation at all, but about reconstructions. The primary origin of the universe is stated in but one verse. This is not the most telling blow against the theory, but it certainly indicates that something has been lost to make the six days of creation anti-climactic.... Or, in the words of Allis: "The first objection to this theory is that it throws the account of creation almost completely out of balance.... It seems highly im-

probable that an original creation which according to this theory brought into existence a world of wondrous beauty would be dismissed with a single sentence and so many verses devoted to what would be in a sense merely a restoration of it" (1954, p. 138, emp. in orig.).

2. The Gap Theory is false because it is based on a forced, artificial, and incorrect distinction between God's creating (*bara*) and making (*asah*). According to the standard rendition of the Gap Theory, these two words **must** mean entirely different things. The term *bara* must refer only to "creating" (i.e., an "original" creation), while the term *asah* can refer only to "making" (i.e., not an original creation, but something "re-made" or "made over"). A review of the use of these two specific Hebrew words throughout the Old Testament, however, clearly indicates that often they are used interchangeably. Morris has commented:

> The Hebrew words for "create" (*bara*) and for "make" (*asah*) are very often used quite interchangeably in Scripture, at least when God is the one referred to as creating or making. Therefore, the fact that *bara* is used only three times in Genesis 1 (vv. 1, 21, and 27) certainly does not imply that the other creative acts, in which "made" or some similar expression is used, were really only acts of restoration. For example, in Genesis 1:21, God "created" the fishes and birds; in 1:25, He "made" the animals and creeping things. In verse 26, God speaks of "making" man in His own image. The next verse states that God "created" man in His own image. No scientific or exegetical ground exists for distinction between the two processes, except perhaps a matter of grammatical emphasis.... Finally, the summary verse (Genesis 2:3) clearly says that **all** of God's works, both of "creating" and "making" were completed with the six days, after which God "rested" (1966, p. 32, emp. in orig.).

The insistence by Gap theorists, and those sympathetic with them, that *bara* **always** must mean "to create something from nothing," is, quite simply, wrong. Such a view has been advocated by such writers as John Clayton[*] (1990a, p. 7) and Hugh Ross[**] (1991, p. 165). Yet Old Testament scholar C.F. Keil, in his commentary, *The Pentateuch,* concluded that when *bara* appears in its basic form, as it does in Genesis 1,

> ...it always means **to create**, and is only applied to a divine creation, the production of that which had no existence before. It is never joined with an accusative of the material, **although it does not exclude a pre-existent material unconditionally**, but is used for the creation of man (ver. 27, ch. v. 1,2), and of everything new that God creates, whether in the kingdom of nature (Numbers xvi.30) or of that of grace (Ex. xxxiv.10; Ps. li.10, etc.) [1971, 1:47, first emp. in orig.; last emp. added].

Furthermore, the Old Testament contains numerous examples which prove, beyond the shadow of a doubt, that *bara* and *asah* **are** used interchangeably. For example, in Psalm 148:1-5, the writer spoke of the "creation" (*bara*) of the angels. But when Nehemiah wrote of that same event, he employed the word *asah* to describe it (9:6). In Genesis 1:1, the text speaks of God "creating" (*bara*) the Earth. Yet, when Nehemiah spoke of that same event, he employed the word *asah* (9:6). When Moses wrote of the "creation" of man, he used *bara* (Genesis 1:27). But one verse before that, he spoke of the "making" (*asah*) of man. Moses also employed the two words

---

[*] The documentation for Clayton's position will be presented in chapter 4 in an examination and refutation of his Modified Gap Theory.

[**] Ross has stated: "The Hebrew word for 'created,' **always** refers to divine activity. The word emphasizes the newness of the created object. It means to bring something entirely new, something previously non-existent, into existence" (1991, p. 165, emp. added).

in the same verse (Genesis 2:4) when he said: "These are the generations of the heavens and of the earth when they were **created** [*bara*], in the day that Jehovah **made** [*asah*] earth and heaven."

Gap theorists teach that the Earth was **created** (*bara*) from nothing in Genesis 1:1. But Moses said in Genesis 2:4 that the Earth was **made** (*asah*). Various gap theorists are on record as stating that the use of *asah* can refer **only** to that which is made from something already in existence. Yet they do not believe that when Moses spoke of the Earth being "made," it was formed from something already in existence.

Consider also Exodus 20:11 in this context. Moses wrote: "For in six days the Lord made [*asah*] heaven and earth, the sea and all that in them is, and rested the seventh day." Gap theorists contend that this verse speaks only of God's "re-forming" from something already in existence. But notice that the verse specifically speaks of the **heaven**, the **Earth**, the **seas**, and **all that in them is**. Gap theorists, however, do not contend that God formed the heavens from something already in existence. The one verse that Gap theorists never have been able to answer is Nehemiah 9:6.

> Thou art Jehovah, even thou alone; thou hast made [*asah*] heaven, the heaven of heavens, with all their host, the earth and all things that are thereon, the seas and all that is in them, and thou preservest them all; and the host of heaven worshippeth thee.

The following quotation from Fields will explain why.

> ...in Nehemiah 9:6 the objects of God's making (*asa*) include the **heavens**, the **host of heavens**, and the **earth**, and **everything contained in and on it**, and the **seas and everything they contain**, as well as the **hosts of heaven** (probably angels).
>
> Now this is a very singular circumstance, for those who argue for the distinctive usage of *asa* throughout Scrip-

ture must, in order to maintain any semblance of consistency, never admit that the same creative acts can be referred to by both the verb *bara* and the verb *asa*. Thus, since Genesis 1:1 says that God **create** (*bara*) the **heavens** and the **earth**, and Exodus 20:11 and Nehemiah 9:6 contend that he **made** (*asa*) them, there must be **two distinct events** in view here. In order to be consistent and at the same time deal with the evidence, gap theorists must postulate a time when God not only "appointed" or "made to appear" the **firmament**, the **sun**, the **moon** and **stars**, and the **beasts**, but there also must have been a time when he only appointed the **heavens**, the **heaven of heavens**, the **angels** (hosts), the **earth, everything on the earth**, the **sea** and **everything in the sea**!

So that, while *asa* is quite happily applied to the firmament, sun, moon, stars, and the beasts, its further application to **everything else contained in the universe**, and, indeed, the universe itself (which the language in both Exodus 20:11 and Nehemiah 9:6 is intended to convey) creates a monstrosity of interpretation which should serve as a reminder to those who try to fit Hebrew words into English molds, that to straitjacket these words is to destroy the possibility of coherent interpretation completely! (1976, pp. 61-62, emp. and parenthetical items in orig.).

Whitcomb was correct when he concluded:

These examples should suffice to show the absurdities to which we are driven by making distinctions which God never intended to make. For the sake of variety and fullness of expression (a basic and extremely helpful characteristic of Hebrew literature), different verbs are used to convey the concept of supernatural creation. It is particularly clear that whatever shade of meaning the rather flexible verb **made** (*asah*) may bear in other contexts of the Old Testament, in the context of **Genesis 1** it stands as a synonym for **created** (*bara*) [1972, p. 129, emp. and parenthetical items in orig.].

3. The Gap Theory is false because there is no justification for translating the verb "was" (Hebrew, *hayetha*) as "became" in Genesis 1:2. Gap theorists insist upon such a translation to promote the idea that the Earth **became** "waste and void" after the Satanic rebellion. Yet usage of the verb *hayah* argues against such a translation. Ramm has noted:

    > The effort to make **was** mean **became** is just as abortive. The Hebrews did not have a word for **became** but the verb **to be** did service for **to be** and **become**. The form of the verb **was** in Genesis 1:2 is the Qal, perfect, third person singular, feminine. A Hebrew concordance will give all the occurrences of that form of the verb. A check in the concordance with reference to the usage of this form of the verb in Genesis reveals that in almost every case the meaning of the verb is simply **was**. Granted in a case or two **was** means **became** but if in the preponderance of instances the word is translated **was**, any effort to make one instance mean **became**, especially if that instance is highly debatable, is very insecure exegesis (1954, p. 139, emp. in orig.).

4. The Gap Theory is false because the words *tohu wabohu* do not mean "something once in a state of repair, but now ruined." Gap theorists believe that God's "initial" creation was **perfect**, but **became** "waste and void" (*tohu wabohu*) as a result of a Satanic rebellion. Whitcomb has addressed this point.

    > Many Bible students, however, are puzzled with the statement in Genesis 1:2 that the Earth was without form and void. Does God create things that have no form and are void? The answer, or course, depends on what those words mean. "Without form and void" translate the Hebrew expression *tohu wabohu*, which literally means "empty and formless." In other words, the Earth was not chaotic, not under a curse of judgment. It was simply empty of living things and without the features that it later possessed, such as oceans and continents, hills and valleys—features that would be essential

for man's well-being. ...when God created the Earth, this was only the first state of a series of stages leading to its completion (1973b, 2:69-70).

5. The Gap Theory is false because there is no evidence to substantiate the claim that Satan's rebellion was on the Earth, much less responsible for a worldwide "cataclysm." The idea of such a cataclysm that destroyed the initial Earth is not supported by an appeal to Scripture, as Morris has explained.

> The great pre-Adamic cataclysm, which is basic to the gap theory, also needs explanation.... The explanation commonly offered is that the cataclysm was caused by Satan's rebellion and fall as described in Isaiah 14:12-15 and Ezekiel 28:11-17. Lucifer—the highest of all God's angelic hierarchy, the anointed cherub who covered the very throne of God—is presumed to have rebelled against God and tried to usurp His dominion. As a result, God expelled him from heaven, and he became Satan, the great adversary. Satan's sin and fall, however, was in heaven on the "holy mountain of God," not on earth. There is, in fact, not a word in Scripture to connect Satan with the earth prior to his rebellion. On the other hand, when he sinned, he was expelled from heaven **to** the earth. The account in Ezekiel says: "Thou wast perfect in thy ways from the day that thou wast created, till iniquity was found in thee. ...therefore I will cast thee as profane out of the mountain of God; and I will destroy thee, O covering cherub, from the midst of the stones of fire. Thine heart was lifted up because of thy beauty, thou has corrupted thy wisdom by reason of thy brightness: I will cast thee to the ground [or 'earth,' the same word in Hebrew]" (Ezekiel 28:15-17).[*]

---

[*] I do not agree with Dr. Morris' comments that Ezekiel 28 and Isaiah 14 refer to Satan. His statements are left intact, however, to show how (even when removed from their proper context) the alleged "proof-texts" used by Gap theorists do not prove a Satanic cataclysm on the Earth. For documentation that Satan is not under discussion in Ezekiel 28 or Isaiah 14, see Thompson, 1998a and 1998b.

There is, therefore, no scriptural reason to connect Satan's fall in heaven with a cataclysm on earth... (1974, pp. 233-234, emp. and bracketed material in orig.).

6. The Gap Theory is false because its most important "proof-text" is premised on a removal of the verse from its proper context. That proof-text is Isaiah 45:18, which reads:

> For thus saith the Lord that created the heavens; God himself that formed the earth and made it; he hath established it, he created it not in vain, he formed it to be inhabited.

In their writings, gap theorists suggest the following. Since Isaiah stated that God did **not** create the Earth *tohu*, and since the Earth of Genesis **was** *tohu*, therefore the latter could not have been the Earth as it was created originally in Genesis 1:1. The implication is that the Earth **became** *tohu* as a result of the cataclysm precipitated by Satan's rebellion against God.

The immediate context, however, has to do with Israel and God's promises to His people. Isaiah reminded his listeners that just as God had a purpose in creating the Earth, so He had a purpose for Israel. Isaiah spoke of God's immense power and special purpose in creation, noting that God created the Earth "to be inhabited"—something accomplished when the Lord created people in His image. In Isaiah 45, the prophet's message is that God, through His power, likewise will accomplish His purpose for His chosen people. Morris has remarked:

> There is no conflict between Isaiah 45:18 and the statement of an initial formless aspect to the created earth in Genesis 1:2. The former can properly be understood as follows: "God created it not (to be forever) without form; He formed it to be inhabited." As described in Genesis 1, He proceeded to bring beauty and structure to the formless elements and then inhabitants to the waiting lands.

It should be remembered that Isaiah 45:18 was written many hundreds of years after Genesis 1:2 and that its context deals with Israel, not a pre-Adamic cataclysm. (1974, p. 241).

7. The Gap Theory is false because it implies death of humankind on the Earth prior to Adam. Pember believed that the fossils (which he felt the Gap Theory explained) revealed death, disease, and ferocity—all tokens of sin. He suggested:

> Since, then, the fossil remains are those of creatures anterior to Adam, and yet show evident token of disease, death, and mutual destruction, they must have belonged to another world, and have a sin-stained history of their own (1876, p. 35).

Pember leveled a serious charge against the Word of God in making such a statement. The idea that the death of humankind occurred **prior** to Adam's sin contradicts New Testament teaching which plainly and emphatically indicates that the death of humankind entered this world as a **result** of Adam's sin (1 Corinthians 15:21; Romans 8:20-22; Romans 5:12). The apostle Paul stated in 1 Corinthians 15:45 that Adam was "the first man." Yet long before Adam—if the Gap Theory is correct—there existed a pre-Adamic race of men and women with (to quote Pember) "a sin-stained history of their own." But how could Paul, by inspiration of God, have written that Adam was the **first** man if, in fact, men had both lived and died before him? The simple fact of the matter is that both the Gap Theory and Paul cannot be correct.

8. The Gap Theory is false because it cannot be reconciled with God's commentary—at the conclusion of His six days of creative activity—that the **whole creation** was "very good."

> Genesis 1:31 records God's estimate of the condition of this world at the end of the sixth day of creation. We read that "God saw **every thing** that he had made, and behold, it was very good. And the evening and the morn-

ing were the sixth day." If, in accordance with the gap theory, the world had already been destroyed, millions of its creatures were buried in fossil formations, and Satan had already become as it were, the god of this world, it is a little difficult to imagine how God could have placed Adam in such a wrecked world, walking over the fossils of creatures that he would never see or exercise dominion over, walking in a world that Satan was already ruling. Could God possibly have declared that everything He had made was very good? In other words, the text of Scripture when carefully compared with this theory creates more problems than the theory actually solves (Whitcomb, 1973b, 2:68-69, emp. added).

9. The Gap Theory is false because of God's plain, simple statement that the Earth and all things in it were made in six days. Wayne Jackson has stated: "The matter can be actually settled by one verse, Exodus 20:11a: 'for in six days Jehovah made heaven and earth, the sea, and **all that in them is**....' If **everything** was made within six days, then **nothing** was created prior to those six days!" (1974, p. 34, emp. in orig.).

In 1948, at the Winona Lake School of Theology. a graduate student, M. Henkel, writing a master's thesis on "Fundamental Christianity and Evolution," polled 20 leading Hebrew scholars in the United States, asking them if there were any exegetical evidences of a gap interpretation of Genesis 1:2. They unanimously replied—No! (Henkel, 1950, p. 49). Nothing has changed in this regard since 1948; the evidence for such a gap still is missing.

4

MISCELLANEOUS
OLD-EARTH THEORIES

Numerous religionists have seen the abject failure of both the Day-Age and Gap theories, yet remain as determined as ever to find a way to force evolutionary time into the biblical text. This misguided determination thus causes them to formulate, modify, temporarily accept, and then abandon theory after theory in search of one they hope eventually will succeed. Unfortunately, many Bible believers have not yet come to the conclusion that the Genesis account is a literal, factual, and defensible record of God's method of creation. And so, rather than accept that account at face value—as Christ and His inspired writers did—they constantly seek some way to alter it by appealing to one theory after another in the hope of ultimately incorporating geologic time into the biblical record. In many instances, the resulting "new" theories are little more than a reworking of the old, discarded theories that long ago were banished to the relic heaps of history because they could not withstand intense examination under the spotlight of God's Word. One such theory making the rounds today is the "Modified Gap Theory." Because of its increasing popularity in certain quarters, I feel it bears examination here.

THE MODIFIED GAP THEORY

Over the past three decades, one of the most frequently used lecturers within certain segments of the churches of Christ has been John N. Clayton, a self-proclaimed former atheist who teaches high school science in South Bend, Indiana. Due to his background in historical geology, Clayton has worked at a feverish pace to produce an amalgamation between the evolutionary geologic record and the Genesis account of creation. Shortly after becoming a Christian, Clayton adopted the position of a full-fledged theistic evolutionist. Later, however, he moved away from strict theistic evolution to an "off-beat" brand of that doctrine that reflects his own "private theology" (see Francella, 1981). Consequently, he is recognized widely by those active in the creation/evolution controversy as the originator and primary defender of what has come to be known as the Modified Gap Theory (see: Clayton, 1976a, pp. 147-148; Thompson, 1977, pp. 188-194; McIver, 1988, 8[3]: 22; Jackson and Thompson, 1992, pp. 115-130).

Since John Clayton advocates that the Earth is roughly 4.5 billion years old (the standard evolutionary estimate), he must accommodate the Genesis account to this concept (see Clayton, 1990b, p. 130). Here, in his own words, is how his Modified Gap Theory attempts to make such an accommodation possible.

> Genesis 1:1 is an undated verse. No time element is given and no details of what the Earth looked like are included. It could have taken place in no time at all, or **God may have used eons of time** to accomplish his objectives. **I suggest that all geological phenomena except the creation of warm-blooded life were accomplished during this time.** There was no way God could have described amoebas, bacteria, viris [sic], or dinosaurs to the ancient Hebrew, and yet these forms of life were vital to the coal, oil and gas God knew man would need. Thus God created these things but did not describe them just as He did not describe a majority of the 110 million species of life on this planet. Changes took place in the Earth (but no gap de-

struction) until God began the formation of man's world with birds, whales, cattle and man in the literal days of Genesis (Clayton, 1976a, pp. 147-148, emp. added).

Clayton has worked on this concept for over thirty years and frequently has altered it in order to make it fit whatever data happen to be in vogue at the time. In his *Does God Exist? Correspondence Course*, he elaborated on what all of this means.

> Not only does the first verse give us the creation of celestial objects, but of a **functional earth** itself.... By the end of Genesis 1:1 there was a functional, living, working earth. If you had stood upon the earth at this point in time, you would have recognized it. Let us once again remind you that how long God chose to use to accomplish this creation is not revealed in this passage.... It is very possible that a living ecosystem operated in Genesis 1:1 to produce the earth. Bacteria may have swarmed in the oceans and giant plants may have lived in great swamps. Dinosaurs may have roamed freely accomplishing their purpose in being. The purpose of all of this would have been to prepare the earth for man. This living ecosystem would have produced the coal, oil, gas, and the like, as well as providing the basis of man's ultimate food supply! (1990c, pp. 3,4, emp. added).

Thus, in capsule form, Clayton suggests that when the Bible says God "created," what it **really** means is that, over eons of time, God "prepared" an Earth for man. Additionally, God did not create **everything** to exist on that "first" Earth. For example, according to the Modified Gap Theory there were no warm-blooded creatures. And, since man is warm-blooded, naturally, he was not there either. Clayton has written:

> I submit to you that Genesis 1:1 is not a summary verse. It is a record of God's action which produced an Earth ready for man's use. I further submit for your consideration that **some time may be involved in this verse and that natural processes may have been used** as well as miraculous ones to prepare the Earth for man (1982, p. 5, emp. added).

Mr. Clayton also has provided an explanation as to **why**, according to his Modified Gap Theory, man was not a part of this original creation.

> The week described in Exodus refers to the week described in Genesis 1:5-31. The week in Genesis 1:5-31 describes the creation of man and a few forms with which man is familiar, but it is **not** a total description of every living thing that does [sic] or ever has existed on Earth (1976b, pp 5-6, emp. in orig.).

Exodus 20:11 explicitly affirms that **everything** that was made by God was completed **within the six days** of the initial week. John Clayton, however, begs to differ with both God and His inspired writer, Moses, and instead asserts that many things actually had been created (during vast epochs of time) long before the creation week ever started. Since, as I already have discussed, Clayton does not believe that Exodus 20:11 refers to **all** of the creative activity of God, but instead refers only to that which occurred in Genesis 1:5-31, he has suggested that Moses "**avoids the creation question** and concentrates on his own purpose" (1976b, p. 5, emp. added). Placed into chart form by Clayton himself, a summary of the Modified Gap Theory appears as you see it below.

CREATION					JUDGMENT
		RESOURCE ECOSYSTEM			
	EARTH SUN	AMOEBA BACTERIA	CHRONOMETRY SET UP		ETERNITY
TIME	MOON	WATER PLANTS	GEN. 1:14-19	GEN. 1:20-28 ADAM NOW	
MATTER	STARS GALAXIES ETC.	DINOSAURS ETC.		CREATION WEEK CHRIST	
				BIRDS WATER MAMMALS MAMMALS MAN	
PRECAMBRIAN	MESOZOIC		CENOZOIC ±		

A Response and Refutation

Whenever John Clayton is challenged regarding the Modified Gap Theory, his usual response is to cloud the issue by suggesting that he does not accept the **standard** Gap Theory. He has said, for example: "You'll notice that I'm accused of advocating both the Gap Theory and the Day-Age Theory there, and of course neither one of those am I advocating.... But I would like to emphasize that I do not in any way, shape, or form embrace the Gap Theory" (1980b). Yet, in his lecture, *Evolution's Proof of God*, he is on record as stating: "In Genesis 1:2 I'm told by the Hebrew scholars that the most accurate reading is that the earth 'became without form and void' and some have suggested that maybe **a tremendous number of years passed between the first part of Genesis 1:1 and Genesis 1:2**" (n.d.[d], emp. added). Mr. Clayton then went on to defend that very position. I wonder—what would the average person call that "tremendous number of years" between Genesis 1:1 and 1:2? A "gap" perhaps?

The "accusation" to which Clayton was responding never suggested that he accepted the **standard** Gap Theory. The issue was whether or not he accepted the **Modified** Gap Theory. [In fact, he is the one who invented the theory in the first place, in the 1976 edition of his book, *The Source*.] The standard Gap Theory suggests that in the alleged time interval between Genesis 1:1 and 1:2, the Earth was destroyed during a battle between Satan and God. Clayton is on record as stating that he does not accept that so-called "gap destruction."

Mr. Clayton does not like being saddled with any label that identifies his views for what they are. He bristles at being "boxed in," to use his own words. In attempting to skirt the issues, sometimes he has been known to answer charges that have not even been leveled. The story surrounding his Modified Gap Theory provides a good example of this very thing.

John Clayton's "private theology" is replete with unscriptural concepts, discrepancies, and contradictions that bear examination. Notice—according to his chart (see page 70)—that the "creation week" does not commence until Genesis 1:14ff. Since this section of Genesis 1 has to do with the events of **day four and afterward**, Clayton's "week" of creative activity has only **three days**! Furthermore, Clayton's Modified Gap Theory suggests that during the eons of time prior to the "creation week" God was **building up** a "resource ecosystem" by the use of amoebas, bacteria, water, plants, dinosaurs, etc. (refer to his chart on Genesis 1:1). Yet at other times, while attempting to defend his Modified Gap Theory, Clayton has contended that the "most accurate reading" of Genesis 1:2 is that the Earth "**became without form and void**" (n.d.[d], emp. added). Which is it? Was the Earth **generating** or **degenerating** during this period? Obviously, it cannot be both.

Earlier, I quoted Clayton as suggesting that in Exodus 20:11 Moses "**avoids the creation question** and concentrates on his own purpose" (1976b, p. 5, emp. added). I would like to address that point here, for it is a careless comment indeed. The purpose of Moses' statement was not merely to **establish** the Sabbath law; it also was an explanation as to the **reason** for the Sabbath. Exactly **why** were the Israelites commanded to observe the Sabbath? Because in six days God created the Earth and its creatures and on the seventh day rested. To state that Moses "avoids the creation question" is wrong. The divine writer did not avoid a reference to the **Creator**; "Jehovah" is specified. Nor did he avoid referring to the Lord's **action**; he noted that God "made" these things.

The Modified Gap Theory flatly contradicts both Exodus 20:11 and Genesis 1. For example, Clayton has argued that the creation of fish (cold-blooded creatures) occurred in Genesis 1:1, whereas according to Moses they were created on the **fifth day**

(Genesis 1:20-23). The Genesis record states that creeping things (which would include both insects and reptiles) were brought into existence on the **sixth day** (1:21,24), but the Modified Gap Theory places them in the time period before the creation week. John Clayton simply rearranges the Genesis record to fit his own evolutionary presuppositions—without any regard whatsoever for what God had to say on the matter.

The only way that Clayton can hold to his Modified Gap Theory and his "private theology" is to convince people that **his way** of translating Genesis is the **correct way**. He has attempted to do just that for more than three decades. In order to succeed, he has found it necessary to present people with an entirely **new vocabulary**. This is the case with many false teachers. They realize they never will be able to reach the masses by using correct, biblical terminology, so they invent altogether new terms, or offer drastic reinterpretations of old ones, in an attempt to make their ideas seem acceptable.

The Modified Gap Theory, with its accompanying off-beat brand of theistic evolution, rests upon the (mis)interpretation of two Hebrew words, *bara* and *asah*, used in Genesis 1-2. Here is what Clayton has said about them, and why they are so important to his theory.

> In the Hebrew culture and in the Hebrew language there is a difference between something being created and something being made. The idea of creation involves a miraculous act on the part of God. It is not something that man can do, nor is it something that can occur naturally.... The Hebrew word used in Genesis 1 to describe this process is the word *bara*. As one might expect, this word is not used extensively in the Bible, in fact, it is only used in verses 1, 21, and 27 in Genesis. The other concept in the Hebrew culture and in the Hebrew language that is used in reference to things coming into existence involves the process of producing something naturally. The idea is that something came into existence because of planning, wisdom, and intelli-

gence, but not as a miraculous act of God. Many times acts of men are described in this way. The Hebrew word *asah* is the main Hebrew word translated this way in Genesis 1. It is vital to a proper understanding of Genesis that these two words not be confused because much understanding is lost and considerable contradiction with the scientific evidence is generated when the words are not distinguished from each other (1991, pp. 6-7).

Clayton also has written:

> We have pointed out that the Hebrew word *bara* normally means to create something out of nothing while the word *asah* usually implies the re-shaping of something that was already in existence.... [T]he normal use of the word *bara* and the normal use of the word *asah* are distinctly different and this difference is important in one's interpretation of Genesis 1 (1979, pp. 2-5).

The following detailed summary from John Clayton's own writings should clarify why this distinction is so important to the success of his Modified Gap Theory.

(1) God initiated the Big Bang, and the Universe developed according to evolutionary theories (Clayton, 1991, p. 8).

(2) The initial creation (*bara*) included such things as the Sun, Moon, Earth, stars, etc. (Clayton, 1991, p. 8). [As I have discussed previously, Clayton puts certain living creatures in this "pre-historic" period (including such things as dinosaurs, bacteria, etc.), but no warm-blooded animals or men.]

(3) Sometime after the initial creation, God then began to form and make (*asah*) things. As Clayton has stated: "It is important to recognize that this process of creating...is described in Genesis 1:1-3. Verse 4ff. deal with something all together different—the making, forming, and shaping of the created earth. Creation does not occur again until animal life is described in verses 20 and 21" (1991, pp. 8-9).

(4) Beginning in the time period called Day 5, according to Clayton, God began to make new things (Clayton, 1991, p. 9), which presumably would **include** marine life, birds, and man, but would **exclude** light, oceans, atmosphere, dry land, planets, stars, moons, and beasts of the field—all of which supposedly were "created" (*bara*) in Genesis 1:1.

(5) Man's spiritual part then was created (*bara*) in God's image (1:27), and his physical part was formed (*yatsar* not *bara*) from the dust of the ground (Clayton, 1991, p. 9).

(6) By the end of Genesis 1, God's "creating" and "making" were completed, but "there is no indication in the Bible that the seventh day ever ended" (Clayton, 1990a, p. 11).

The convoluted scenario involved in what you have just read is necessary, from Clayton's viewpoint, in order to make his Modified Gap Theory work. Here, now, is what is wrong with all of this. First, the distinction of the alleged difference between *bara* and *asah* is as artificial as it is as stilted, and Clayton has admitted as much. In the May 1979 issue of the *Does God Exist?* journal that he edits, he wrote:

> **Because there are a few isolated exceptions** where the context seems to indicate that the word *bara* or *asah* has been used in a different way than the application we have just discussed, there are those who maintain that one cannot scripturally maintain the applications of these words as we have presented them in reference to Genesis 1. The Hebrew language, as most of us recognize, is a language which can be interpreted only in its context (1979, p. 4, emp. added).

In his *Does God Exist? Correspondence Course*, Clayton also confessed: "Some may object to this superliteral interpretation of *bara* and *asah* by responding that there are exceptions to the usages I have described in the previous paragraphs. **Such a criticism is valid**" (1990d, p. 3, emp. added).

Second, the "few isolated exceptions," as Clayton calls them, are neither few nor isolated. Furthermore, they obliterate his artificial distinction in regard to *bara* and *asah* (which often are used **interchangeably** throughout the Old Testament and do not always have the strict interpretation that Clayton has placed on them). Notice the following.

(1) Clayton has suggested: "As one might expect, this word [*bara*—BT] is not used extensively in the Bible, in fact, it is only used in verses 1, 21, and 27 in Genesis." This statement is completely untrue. *Strong's Exhaustive Concordance* cites no fewer than 11 instances of *bara* in the book of Genesis. Additionally, *bara* and its derivatives occur 40 times in the Old Testament apart from Genesis. In over 30 instances, it means "create, shape, form, or fashion."

(2) Clayton has insisted—according to the rules of his "new vocabulary"—that the word *bara* **always** must mean "to create something from nothing" (1990a, p. 7). This, too, is incorrect, as I argued at length in chapter 3. Henry Morris observed:

> The Hebrew words for "create" (*bara*) and for "make" (*asah*) are very often used quite interchangeably in Scripture, at least when God is the one referred to as creating or making. Therefore, the fact that *bara* is used only three times in Genesis 1 (vv. 1, 21, and 27) certainly does not imply that the other creative acts, in which "made" or some similar expression is used, were really only acts of restoration. For example, in Genesis 1:21, God "created" the fishes and birds; in 1:25, He "made" the animals and creeping things. In verse 26, God speaks of "making" man in His own image. The next verse states that God "created" man in His own image. No scientific or exegetical ground exists for distinction between the two processes, except perhaps a matter of grammatical emphasis.... Finally, the summary verse (Genesis 2:3) clearly says that **all** of God's works, both of "creating" and "making" were completed with the six days, after which God "rested" (1966, p. 32, emp. in orig.).

Clayton's insistence that *bara* **always** must mean "to create something from nothing, is, quite simply, wrong. Old Testament scholar C.F. Keil concluded that when *bara* appears in its basic form, as it does in Genesis 1,

> ...it always means **to create**, and is only applied to a divine creation, the production of that which had no existence before. It is never joined with an accusative of the material, **although it does not exclude a pre-existent material unconditionally**, but is used for the creation of man (ver. 27, ch. v. 1,2), and of everything new that God creates, whether in the kingdom of nature (Numbers xvi. 30) or of that of grace (Ex. xxxiv.10; Ps. li.10, etc.) [1971, 1:47, first emp. in orig.; last emp. added].

Furthermore, there is ample available evidence that John Clayton knows his efforts to make *bara* represent **only** that "which has been created from nothing" are incorrect. Genesis 1:27 is the passage that reveals the error of his interpretation: "So God created (*bara*) man in his own image, in the image of God created he him; male and female created he them." If Clayton is correct in his assertion that *bara* can be used **only** to describe something created from nothing, then the obvious conclusion is that in Genesis 1:27 God created man and woman **from nothing**. But such a view conflicts with Genesis 2:7, which states that God formed man from the dust of the ground.

How has Clayton attempted to correct his obvious error? He has suggested—in keeping with his new vocabulary—that Genesis 1:27 **really** is saying that when God "created" (*bara*) man, He actually created not man's body, but his **soul** from nothing (1991, p. 9). Such a strained interpretation can be proven wrong by a simple reading of the text. Genesis 1:27 states **what** was created—"**male** and **female** created he them." The question then must be asked: Do souls come in "male" and "female" varieties? They do not. Souls are spirits, and as such are sexless, (e.g., as Jesus said angels were—Matthew 22:29-30). Yet the Modified Gap

Theory plainly implies that male and female souls exist. A well-known principle in elementary logic is that any argument with a false premise (or false premises) is unsound. Thus, the Modified Gap Theory is unsound.

(3) Taking the creation passages at face value and in their proper context, it is obvious that no distinction is made between the act of creating and the act of making. For example, God's activity during this first week is described in terms other than creating or making. This includes the phrase, "Let there be," which is used to usher in each new day and the things created in that day. Also, note that God "divided" the light from the darkness, and He "set" the light-giving objects in the expanse of the sky. How would John Clayton's new vocabulary deal with these verbs?

(4) There is compelling evidence that the words *bara* and *asah* are used **interchangeably** throughout the Old Testament. Mr. Clayton, of course, adamantly denies that this is the case. He has stated: "It is difficult to believe that there would be two words used to convey the same process" (1990a, p. 7). Yet why is it difficult to imagine that two different words might be used to describe exactly the same process? Writers commonly employ different words to describe the same thing(s), thereby providing "stylistic relief"—a grammatical construct which avoids the needless repetition that occurs by using the same words over and over. For more than a hundred years, conservative scholars have made a similar point to proponents of the Documentary Hypothesis, arguing that there is no reasonable way to "dissect" the Old Testament on the basis of the words *Elohim* (translated "God") and Yahweh (translated "Jehovah" or "Lord").

Bible writers often employed different words to describe the same thing(s). For example, in the four Gospels Christ is spoken of as having been killed, crucified, and slain. Where is the distinction? New Testament writers often spoke of the church, the body, and the kingdom—which are exactly the same thing.

Where is the difference? Why should anyone find it so difficult to accept that different words may be used to describe the same thing or event?

Furthermore, the Scriptures are replete with examples which prove that *bara* and *asah* **are** used interchangeably. For example, in Psalm 148:1-5 the writer spoke of the "creation" (*bara*) of the angels. Yet when Nehemiah addressed the creation of angels, he employed the word *asah* to describe it (9:6). In Genesis 1:1, as Clayton has admitted, the text speaks of God "creating" (*bara*) the Earth. Yet again, when Nehemiah spoke of that same event, he employed the word *asah* to do so (9:6). When Moses wrote of the "creation" of man, he used *bara* (Genesis 1:27). But one verse before that (1:26), he spoke of "making" (*asah*) man. Moses also employed the two words **in the same verse** (Genesis 2:4): "These are the generations of the heavens and of the earth when they were **created** [*bara*], in the day that Jehovah **made** [*asah*] earth and heaven."

John Clayton has said that the Earth was **created** (*bara*) from nothing in Genesis 1:1. Yet Moses said in Genesis 2:4 that the Earth was **made** (*asah*). Clayton has stated that the use of *asah* can refer **only** to that which is made from something already in existence. Does he believe that when Moses spoke of the Earth being "made," it was formed from something already in existence?

And what about Exodus 20:11 in this context? Moses wrote: "For in six days the Lord made [*asah*] heaven and earth, the sea and all that in them is, and rested the seventh day." Clayton has written that this speaks only of God's "forming" from something already in existence. But notice that the verse specifically speaks of the **heaven**, the **Earth**, the **seas**, and **all that in them is**. Does Clayton therefore contend that God formed the heavens from something already in existence? Exodus 20:11 speaks of **everything** made by God in the six days of creation. Yet even he has admitted that "creation (*bara*) does not occur again until animal life is described in verses 20 and 21." How can this be? Moses stated that God "made" (*asah*) everything in the creation week. Now Clayton says there was "creation" (*bara*) going on in that same week. Even John Clayton, therefore, has admitted that there are times when the two words describe the same events during the same time period!

In addition to these problems, the Modified Gap Theory has the same difficulty explaining Nehemiah 9:6 as the standard Gap Theory. Since that was discussed in Chapter 3, I will not be repeat it here.

(5) Weston W. Fields suggested that forcing *bara* and *asah* to refer to completely separate acts results in a "monstrosity of interpretation"—which is exactly what John Clayton's suggested usage of these words represents. Remember that Clayton teaches that at the end of Genesis 1:1, there existed a **fully functional** Earth (complete with various kinds of life teeming on it) and that it remained that way for eons of time. If that is the case—based on his *bara/asah* argument—how would he explain the following problem?

Clayton has taught that the "heavenly bodies" (Sun, Moon, stars, etc.) were a part of the *bara*-type creation of Genesis 1:1. But Exodus 20:11 specifically states that they were "made" (*asah*). Are we to believe that they were **both** "created" **and** "made"? Yes, that is exactly what Clayton has advocated.

> Applied in this literal sense to Genesis 1, one would find that the **heaven and earth** were brought into existence miraculously in Genesis 1:1. This would include the sun, moon, stars, galaxies, black holes, nebula, comets, asteroids and planets.... Verses 14-19 would not describe the creation of the sun, moon and stars, but the reshaping or rearranging of them to a finished form (1989, p. 6).

How were the Sun, Moon, and stars ("created," Clayton says, in Genesis 1:1) assisting the Earth in being "fully functional" when they themselves had not even been "rearranged to a finished form"? One hardly could have a fully functional Earth without the Sun and Moon. Yet by his own admission, Genesis 1: 14-19 speaks of God doing **something** to those heavenly bodies. For centuries Bible scholars have accepted that it is in these verses that God is described as bringing the heavenly bodies into existence. But no, says Clayton, that is not true. They were in existence from Genesis 1:1, but they had not yet been "rearranged to a finished form"—something that would not occur until billions of years later. How could these **unfinished** heavenly bodies have been of any use to a **finished** Earth? How could the Earth be "functional" unless the Sun, Moon, and other planets were "functional" as well? And if they were "functional" in Genesis 1:1, why "rearrange" them?

Clayton is on record as stating: "When we look at those places where the word 'make' is used, the context leaves absolutely no doubt about what the intention of the author is for that passage" (1979, p. 5). I could not agree more. There is **absolutely no doubt** about how the Bible writers employed these words. They used them just as any author would employ them—interchangeably.

THE NON-WORLD VIEW OF ORIGINS

Imagine, if you will, the dilemma of a person who has done all he knows to do to force the evolutionary geologic age-system into the biblical record, but who has discovered that it simply will not fit. If that person wishes to retain his belief in God, but abjectly refuses to accept the biblical account of creation at face value, what option remains open to him? For the person who does not wish to become an atheist or agnostic, there is only one remaining possibility—the so-called "Non-World View."

The Non-World View dates from 1972 with the publication of *A Christian View of Origins* by Donald England of Harding University. In essence, it is a "refusal to get involved" by not taking a stand on the Genesis account of creation. England himself defined it as suggesting:

There is no world view presented in Genesis 1. I believe the intent of Genesis 1 is far too sublime and spiritual for one to presume that it teaches anything at all about a cosmological world view. We do this profound text a great injustice by insisting that there is inherent within the text an argument for any particular world view (1972, p. 124).

In other words, this is a compromise for the person who: (a) refuses to accept the Genesis account of creation as written; but (b) cannot find a reasonable alternative. In his book, Dr. England made it clear that from a straightforward reading of the Genesis account "one gets the general impression from the Bible that the earth is young," and that "it is true that Biblical chronology leaves one with the general impression of a relatively recent origin for man" (1972, p. 109). But he also made it clear that he had no intention of accepting such positions, since they disagree with "science."

Finding himself painted into a corner, as it were, the only way out was simply to throw up his hands and, with a sigh of relief, view Genesis as containing **no world view whatsoever**. As one writer who strongly recommends the Non-World View has suggested:

By "Non-World" we mean that we don't accept any "God-limiting" position on how we interpret Genesis. We don't limit our comprehension of time, space, or process in any way Biblically; and do this unlimiting on the basis that that's what God intended....

If Chapter 1 is not a detailed historical account, how do we fit the fossil record to it? The "Non-World" View says "we don't." If we are to speak where the Bible speaks and be silent where the Bible is silent we won't succumb to the pressure to make it fit. Since the Bible doesn't mention dinosaurs, bats, amoeba, bacteria, DNA, virus [sic], sea plants, algae, fungus [sic], etc., we won't attempt to match them. There are a few forms we can match, but only a few out of the millions. The Hebrew words used in Genesis do not cover whole phyla of animals but they are reasonably specific. If we take a "Non-World View," this doesn't bother us because we are only interested in God's message to man, not in satisfying man's curiosity.

The "Non-World View" also finds no necessity in dealing with men's arguments on the scientific theories of creation and age. There is no necessity to argue about the "big bang," "steady state," or irtron theory of origins; nor is there any need to hassle about whether the Earth is 6, 6,000 or 6 billion years old. Genesis 1:1 says only that God did it! That is the purpose. It is **not** the purpose to state how or when (Clayton, 1977, pp. 6-8, emp. in orig.).

A Response and Refutation

The careful reader soon will realize that this is indeed the compromise to end all compromises. With the Non-World View, a person may believe as much, or as little, as he wants in regard to the Genesis account of creation. If the person who holds to this view is challenged with a passage of Scripture, he may reply simply, "Oh, that passage doesn't have any particular world view in it." And the convenient thing is that it does not matter how forceful the passage may be, whether it comes from the Old Testament or the New, what biblical writer may have penned it, or even if Christ Himself spoke it. With the Non-World View, **everything** is subjective.

The beauty of this view, according to Clayton, is that it is not "God-limiting" (1977, p. 6). Even though when one reads the creation account he definitely gets the "general impression" that man has been here only a short while, and that the Earth is relatively young, and even though Christ Himself stated in Mark 10: 6 that man and woman have been here "from the beginning of the creation," all of that becomes irrelevant. With a wave of the hand, Genesis 1 means little-to-nothing. In fact, it might as well not have been written, for it simply has "no world view" in it at all.

Yet God went to great lengths to explain what was done on day one, what was done on day two, and so on. He told Moses that He took six days to do it. Then He set the Sabbath day as the Jews' remembrance of His creative acts on those days. If God said "in the beginning" and "in six days the Lord created," that is a **time element**. Jesus Himself said that, "**from the beginning of the creation**, male and female made He them" (Mark 10:6). That is a time element. While it may not give an exact day and hour, it says much. It says man was on the Earth "from the beginning." That automatically rules out an ancient Earth, and those compromising theories intent on having one (e.g., the Day-Age Theory, Gap Theory, Modified Gap Theory, etc.). God has indicated, in a way we can understand, what He wants us to know about the time element. When He wrote that He created "the heavens, the earth, the seas, and all that in them is" in six days, does that sound like a Non-World View?

Man may not understand completely the "how" of creation, but it is present nevertheless. When the Scriptures say, "And God said, 'Let there be light' and there was light"—that is **how**. When the Scriptures say, "And God said, 'Let the earth put forth grass',", and later "And the earth brought forth grass"—that is **how**. The "how" is by the power of God (cf. Hebrews 1:3, wherein the writer declared that God upholds "all things by the word of his power").

Granted, the text of Genesis 1 is sublime and spiritual. **It also is historical**. Jesus Christ Himself said so (Matthew 19:4). So did Paul (1 Corinthians 15:45; Romans 8:22; 1 Timothy 2:13). That should settle the matter. God said **that** He did it. God said **how** He did it—"by the word of his power." God said **when** He did it—"in the beginning." The honest reader eventually will come to realize just how much that **in**cludes, and just how much it **ex**cludes. The only "world view" left is the perfect one—that of Genesis 1.

The Non-World View is a neatly disguised but openly flagrant attack on Genesis 1. It impeaches the testimony of the New Testament writers and impugns the integrity of the Lord Himself. And for what purpose? What ultimate good does it accomplish? It merely compromises the truth, while leaving open the way for any and all viewpoints on creation—whether founded in Scripture or not. Furthermore, surely the question needs to be asked: **If Genesis 1 is not God's world view, what is?**

THE MULTIPLE GAP THEORY

For those who find the Day-Age and Gap theories impossible to defend, and yet who do not wish to opt for a theory like the Non-World View that is an open door to extreme liberalism and/or modernism, the list of remaining available theories is quite short. One concept that has become somewhat popular is known as the "Multiple Gap Theory."

The Multiple Gap Theory suggests that the creation days were, in fact, six literal, 24-hour days during which God actually performed the special creative works attributed to Him in Genesis 1. However, these literal days tell only a small part of the whole story. Rather than representing the totality of God's work in creation, they instead represent "breaks" between the geologic ages. In other words, after God's activity on any given literal day, that day then was followed by long ages of slow development in the style of orthodox historical geology. Actually, this theory is a hybridization of the Day-Age and Gap theories. Instead of making "ages" out of the days of Genesis 1, it merely inserts the ages **between** the days. And instead of putting a gap in between Genesis 1:1 and 1:2, it inserts gaps between the days of Genesis 1.

One of the Multiple Gap Theory's strongest supporters, and certainly one of its most ardent popularizers, is, strangely enough, Donald England. The reason I say "strangely enough" is because this is the same Donald England, quoted above, who invented the Non-World View of Genesis and who is on record as stating:

Genesis 1 is far too sublime and spiritual for one to presume that it teaches anything at all about a cosmological
world view. We do this profound text a great injustice by insisting that there is inherent in the text an argument for any
particular world view (1972, p. 124).

Of course, as I already have pointed out, the main reason for
postulating the Non-World View of Genesis 1 is so that a person
may insert into the text **any** world view that he happens to hold
at any given moment. That is exactly what has happened in the
case of Dr. England and the Multiple Gap Theory. A word of explanation is in order.

Dr. England, as I noted earlier, is a professor at Harding University. Harding is supported by members of the churches of Christ,
who generally are known to be quite conservative in their positions regarding the Genesis account of creation. In the past, for
the most part, members of the churches of Christ have not toler-
ated the teachings of false doctrines associated with creation. Dr.
England, of course, is well aware of that fact. The Multiple Gap Theory has the advantage of allowing him, when asked, to assert that
he does, in fact, believe the days of creation to be 24-hour periods.
And, if he is asked if he believes in the Gap Theory, again, he can
demur, insisting that he does not.

But is this an upright approach? Or is it "playing loosely with
the facts"? Interestingly, an example is available upon which one
may base an answer to such questions. In March 1982, Dr. England lectured to a group of young people in Memphis, Tennes-
see. During that series, he told these youngsters that although he
had spent a lifetime searching for "proof" that the days of Genesis 1 were 24-hour days, he never had found any. He then went
to great lengths to set before this audience of impressionable
teenagers a number of "objections" to the days of Genesis 1 being literal days.

As a result of Dr. England's comments, and a subsequent review of them (see Thompson, 1982), the then-President of Har-

ding University, Clifton L. Ganus, received several inquiries from the school's financial supporters about Dr. England's position on these matters. How did England respond? On October 4, 1982 he wrote Dr. Ganus a letter in which he stated:

> Dear Dr. Ganus: I enjoyed my brief visit with you on Friday afternoon. I stated in your presence that I have always believed that the creation days of Genesis One were six twenty-four hour days. Anyone who would take anything that I said in the [name of congregation omitted here—BT] lectures and try to associate me with a "day-age" theory of creation is making a mistake.... Whenever I speak on the creation theme, I am always careful to make my position clear as to my understanding of the length of days in Genesis One.... (1982, p. 1).

England then offered, as proof of his position on these matters, a quotation from pages 111-113 of his book, *A Christian View of Origins*, in which he wrote that he does not recommend strict theistic evolution. But here is the interesting point in all of this. In that same book, just two pages earlier, Dr. England had written:

> The statements, "God created" (Genesis 1 and elsewhere) and "God spoke and it was done; He commanded and it stood fast" (Ps. 33:9) do not explicitly rule out some sort of process. Now, if the days of Genesis are taken as 24-hour days, then that certainly rules out any process extending over vast periods of time. The days could easily have been twenty-four-hour days and the earth still date to great antiquity **provided that indefinite periods of time separated the six creation days** (1972, pp. 110-111, emp. added).

Is this dealing honestly with the facts? Dr. England told the university president (who certainly had the power to dismiss him from his professorial position) that he **does believe** the days of Genesis 1 were 24-hours long, all the while knowing that he has defended, in print, the Multiple Gap Theory.

A Response and Refutation

At the very least, this theory requires a most "unnatural" reading of the Creation account which apparently is continuous and meant to describe the creation of "heaven and earth, the sea, and all that in them is." The context of the creation record suggests continuity. There is absolutely no exegetical evidence to document the claim that in between each of the (literal) creation days there were millions or billions of years. In fact, such evidence is conspicuously missing. In his 1983 volume, *A Scientist Examines Faith and Evidence*, Dr. England commented on this fact when he said: "True, the silence of the Scriptures leaves open the possibility of time gaps but **it does not seem advisable to build a doctrinal theory on the basis of a silence of Scripture**" (p. 154, emp. added).

Nor does the theory harmonize with orthodox geology. If the acts of creation are left on their respective days, then there is no possible way to make the Creation account agree with the geologic-age system—gaps or no gaps. As the chart on the next page shows, the Genesis sequence and the alleged geologic sequence **do not agree**. The Multiple Gap Theory does not alter that fact.

Additionally, we must not overlook the import of Exodus 20:11 which states that "in six days the Lord made heaven and earth, the sea, and **all** that in them is, and rested on the seventh day." Either God made what He made in six days or He made what He made in six days **plus** millions or billions of years. Those who respect the Bible as the inspired Word of God have no trouble accepting the former and rejecting the latter. It is fitting that we close this chapter with a quotation from G. Richard Culp:

> We stand either with God and His teaching of creation, or we stand with the evolutionist in opposition to Him. The issues are sharply drawn; there can be no compromise. You are either a Christian or an evolutionist; you cannot be both. God wants prophets, not politicians; not diplomats, but soldiers in the spiritual sense (1975, p. 163).

DIVISION	SIGNIFICANT FOSSIL APPEARANCES	YEARS AGO (millions)	GENESIS (days)
Cenozoic			
Quaternary	*Homo erectus/H. sapiens*	2	6
Tertiary	Rabbits; Rodents; Marsupials		6
	Camels; Deer; Cattle; Horses		6
	Elephants; Pigs; Early marsupials		6
	Whales; Dolphins; Seals	65	5
Mesozoic			
Cretaceous	Flowering plants		3
	Platypus; Sloths		6
	Modern bony fishes		5
	Snakes	144	6
Jurassic	Lizards		6
	Birds		5
	First true mammals	208	6
Triassic	Turtles; Frogs; Crocodiles		5
	Tuatara; Dinosaurs		6
	Conifers	245	3
Paleozoic			
Permian	Ginkgoes; Cycads; Horsetails		3
	Marine reptiles	286	5
Carboniferous	Reptiles; Mammal-like reptiles		6
	Amphibians		5
	Ferns	360	3
Devonian	Sharks; Bony fish	408	5
Silurian	Club mosses	438	3
Ordovician	Jawless fishes	505	5
Cambrian	Worms; Shellfish; Trilobites		5
	Burgess Shale fauna; First fish?	550	5
Precambrian			
Proterozoic	Jellyfish; Ediacaran fauna		5
	Green algae	2,500	3?
Archaean	Bacteria	3,800	2?
Hadean	First single-celled organism		
	Formation of Earth and Moon		2-4
	Formation of Solar System	4,800	4

Comparison of the evolutionary geological column with the order of creation in Genesis. Evolutionary dates and data based primarily on Gould (1993).

5

BIBLICAL GENEALOGIES AND THE AGE OF THE EARTH

Attempts to place the time necessary for an ancient Earth **during** the creation week (i.e., the Day-Age Theory) have proven unsuccessful. Similarly, attempts to insert the time necessary for an old Earth **before** the creation week (i.e., the Gap Theory) also have failed. Subsequently, the suggestion has been made that perhaps geologic time might be placed **after** the creation week of Genesis 1.

Those willing to offer such a suggestion, however, have been few and far between because of a major obstacle in the biblical record to such a compromise. As every student of the Sacred Scriptures is aware, the Bible contains lengthy genealogies. That these records play a vital role in biblical literature is clear from the amount of space devoted to them in God's Word. Furthermore, they provide a tremendous protection of the text via the message they tell. That message is this: **man has been on the Earth since the beginning, and that beginning was not very long ago**.

While genealogies (and chronologies) serve various functions in the literature of Scripture, one of their main purposes is to show the historical connection of great men to the unfolding of Jehovah's redemptive plan. These lists, therefore, form a connecting link from the earliest days of humanity to the completion of God's salvation system. In order for them to have any evidential value, the lists must be substantially complete.

In the introduction to this book, I made the point that the inspired writer of Hebrews, in contending for the heavenly nature of Christ's priesthood, argued that the Savior could not have functioned as a priest while He was living upon the Earth since God had a **levitical** priesthood to accomplish that need (Hebrews 8:4). Jesus did not qualify for the levitical priesthood for "it is **evident** that our Lord hath sprung out of **Judah**" (Hebrews 7:14, emp. added). I then asked: How could it have been "evident" that Jesus Christ was from the tribe of Judah unless there were accurate genealogical records by which such a statement could be verified? The writer of Hebrews based his argument on the fact that the readers of his epistle would not be able to dispute the ancestry of Christ due to the reliable nature of the Jewish documentation available—i.e., the genealogies.

Yet some Bible believers—determined to incorporate evolutionary dating schemes into God's Word—have complained that the biblical genealogies may not be used for chronological purposes because they allegedly contain huge "gaps" that render them ineffective for that purpose. Donald England has suggested, for example: "Furthermore, it is a misuse of Biblical genealogies to attempt to date the origin of man by genealogy" (1983, p. 155). John Clayton advocated the same view when he wrote: "Any attempt to ascribe a specific or even a general age to either man or the Earth from a Biblical standpoint is a grievous error" (n.d.[a], p. 3). Clayton also stated: "The time of man's beginning is not even hinted at in the Bible. There is no possible way of determining when Adam was created" (n.d.[b], p. 2).

In so commenting, most writers reference (as does Clayton in his writings) the nineteenth-century author, William H. Green (1890), whose writings on the genealogies usually are accepted uncritically—and acclaimed unjustifiably—by those who wish to insert "gaps" (of whatever size) into the biblical genealogies. Thus, we are asked to believe that the genealogies are relatively useless in matters of chronology.

However, these same writers conspicuously avoid any examination of more recent material which has shown that certain portions of Green's work either were incomplete or inaccurate. And while references to the genealogies of Genesis 5 and 11 are commonplace, discussions of material from chapter 3 of Luke's Gospel appear to be quite rare. Two important points bear mentioning in regard to genealogical listings. First, to use the words of Custance:

> We are told again and again that some of these genealogies contain gaps: but what is never pointed out by those who lay the emphasis on these gaps, is that they only know of the existence of these gaps because the Bible elsewhere fills them in. How otherwise could one know of them? But if they are filled in, they are not gaps at all! Thus, in the final analysis the argument is completely without foundation (1967, p. 3).

If anyone wanted to find "gaps" in the genealogies, it was Dr. Custance—who spent his entire adult life searching for a way to accommodate the Bible to an old-Earth scenario. Yet even he admitted that arguments alleging that the genealogies contain sizable gaps are unfounded.

Second—and this point cannot be over-emphasized—**even if there were gaps in the genealogies, there would not necessarily be gaps in the chronologies therein recorded. The question of chronology is not the same as that of genealogy**. This is a critical point that has been overlooked by those who suggest that the genealogies are "useless" in matters of chronology. The "more recent work" mentioned above that documents the accuracy of the genealogies is from James B. Jordan, who reviewed Green's work and showed a number of his arguments to be untrustworthy. To quote Jordan:

> Gaps in genealogies, however, do not prove gaps in chronologies. The known gaps all occur in non-chronological genealogies. Moreover, even if there were gaps in the genealogies of Genesis 5 and 11, this would not affect the chronological information therein recorded, for even if Enosh were the great-grandson of Seth, it would still be the case that Seth was 105 years old when Enosh was born, according to a simple reading of the text. Thus, genealogy and chronology are distinct problems with distinct characteristics. They ought not to be confused (1979/1980, p. 12).

Unfortunately, many who attempt to defend the concept of an ancient Earth **have** confused these two issues. For example, some have suggested that abridgment of the genealogies has occurred and that these genealogies therefore cannot be chronologies, when, in fact, exactly the opposite is true—as Jordan's work has documented. Matthew, as an illustration, was at liberty to arrange his genealogy of Christ in three groups of 14 (making some omissions) because his genealogy was derived from more complete lists found in the Old Testament. In the genealogies of Genesis 5 and 11, remember also that the inclusion of the father's age at the time of his son's birth is wholly without meaning unless chronology was intended. Else why would the Holy Spirit have provided such "irrelevant" information?

Unfortunately, there can be little doubt that some have painted an extremely distorted picture for audiences (or readers) by suggesting that substantial "gaps" occur in the biblical genealogies. Such a distorted picture results, for example, when it is suggested that genealogy and chronology are one and the same problem, for they most certainly are not.

Plus, there are other important considerations. Observe the following information in chart form. Speaking in round figures, from the present to Jesus was roughly 2,000 years—a figure obtainable via secular, historical documents. From Jesus to Abraham also was approximately 2,000 years—another figure that is verifiable historically.

Present to Jesus	2,000 years
Jesus to Abraham	2,000 years
Abraham to Adam	? years

The only figure missing is the one that represents the date from Abraham to Adam. Since we know that Adam was the first man (1 Corinthians 15:45), and since we know that man has been on the Earth "from the beginning of the creation" (Mark 10:6; cf. Romans 1:20-21), if it were possible to obtain the figures for the length of time between Abraham to Adam we then would have chronological information providing the relative age of the Earth (since we also know that the Earth is only five days older than man—Genesis 1; Exodus 20:11; 31:17).

The figure representing the time span between Abraham and Adam, of course, is **not** obtainable from secular history (nor should we expect it to be) since large portions of those records were destroyed in the Great Flood. But the figure is obtainable—via the biblical record. Allow me to explain.

First, few today would deny that from the present to Jesus has been approximately 2,000 years. [For our purpose here, it does not matter whether Christ is viewed as the Son of God since the discussion centers solely on the fact of His existence—something that secular history documents beyond doubt.] Second, in Luke 3 the learned physician provided a genealogy that encompassed 55 generations spanning the distance between Jesus and Abraham—a time frame that archaeology has shown covered roughly 2,000 years (see Kitchen and Douglas, 1982, p. 189). Third, Luke documents that between Abraham and Adam there were only twenty generations. Thus, the chart now looks like this:

Present to Jesus	2,000 years
Jesus to Abraham	2,000 years (55 generations)
Abraham to Adam	? years (20 generations)[*]

Since Genesis 5 provides the ages of the fathers at the time of the births of the sons between Abraham and Adam (thus providing chronological data), it becomes a simple matter to determine the approximate number of years involved. In round numbers, that figure is 2,000. The chart then appears as follows.

Present to Jesus	2,000 years
Jesus to Abraham	2,000 years (55 generations)
Abraham to Adam	2,000 years (20 generations)

Of course, some have argued that there are "gaps" in the genealogies (e.g., Clayton, 1980a, pp. 6-7). But where, exactly, should those gaps be placed, and how would they help? Observe the following. No one can place gaps between the present and the Lord's birth because secular history accurately records that age-information. No one can insert gaps between the Lord's birth and Abraham because secular history also accurately records that age-information. The only place one could put any "usable" gaps (viz., usable in regard to extending the age of the Earth) would be in the 20 generations between Abraham and Adam. Yet notice that there are not actually 20 generations available for the insertion of gaps because Jude stated that "Enoch was the **seventh**

---

* The reader may wonder how 55 generations (Jesus to Abraham) could cover 2,000 years while 20 generations (Abraham to Adam) also cover 2,000 years. The answer, of course, lies in the ages of the patriarchs. Because they lived to such vast ages, fewer generations were required to encompass the same number of years. For an in-depth discussion of the Bible, science, and the ages of the patriarchs, see Thompson, 1995, pp. 265-275.

from Adam" (Jude 14). An examination of the Old Testament genealogies establishes the veracity of Jude's statement since, counting from Adam, Enoch **was** the seventh. Jude's comment thus provides divinely inspired testimony regarding the accuracy of the first seven names in Luke's genealogy—thereby leaving only 13 generations into which any gaps could be placed.

In a fascinating article some years ago, Wayne Jackson observed that in order to accommodate the biblical record only as far back as the appearance of man's alleged evolutionary ancestor (approximately 3.6 million years), one would have to place **291,125 years** between **each** of the remaining 13 generations (1978, 14[18]:1). It does not take an overdose of either biblical knowledge or common sense to see that this quickly becomes ludicrous in the extreme for two reasons. First, who could believe that the first seven of these generations are so **exact**—while the last thirteen are so **inexact**? Is it proper biblical exegesis to suggest that the first seven listings are correct as written, but gaps covering more than a quarter of a million years may be inserted between each of the last thirteen? Second, what good would all of this do anyone? All it would accomplish is the establishment of a 3.6 **million** year-old Earth; old-Earth creationists, progressive creationists, and theistic evolutionists need a 4.6-**billion**-year-old Earth. So, in effect, all of this insertion of "gaps" into the biblical text is much ado about nothing.

And therein lies the point. While it may be true on the one hand to say that an **exact** age of the Earth is unobtainable from the information contained within the genealogies, at the same time it is important to note that—using the best information available to us from Scripture—the genealogies hardly can be extended to anything much beyond 6,000 to 7,000 years. For someone to suggest that the genealogies do not contain legitimate chronological information, or that the genealogies somehow are so full of gaps as to render them useless, is to misrepresent the case and distort the facts.

6

ADDITIONAL CONSIDERATIONS

In any discussion of the Bible and the age of the Earth, there are several additional considerations that should be examined as well.

"FROM THE BEGINNING OF THE CREATION"/ "FROM THE CREATION OF THE WORLD"

In Mark 10:6, Jesus declared concerning Adam and Eve: "**But from the beginning of the creation**, Male and female made he them" (cf. Matthew 19:4). Christ thus dated the first humans from the creation week. The Greek word for "beginning" is *arché*, and is used of "**absolute**, denoting **the beginning of the world** and of its history, the beginning of creation." The word in the Greek for "creation" is *ktiseos*, and denotes "**the sum-total of what God has created**" (Cremer, 1962, pp. 113,114,381, emp. in orig.).

Bloomfield noted that "creation" in Mark 10:6 "signifies 'the things created,' the world or universe" (1837, 1:197-198). In addressing this point, Wayne Jackson wrote:

Unquestionably this language puts humankind at the very dawn of creation. To reject this clear truth, one must contend that: (a) Christ knew the Universe was in existence billions of years prior to man, but accommodating Himself to the ignorance of His generation, deliberately misrepresented the situation; or, (b) The Lord, living in pre-scientific times, was uninformed about the matter (despite the fact that He was there as Creator—John 1:3; Colossians 1:16). Either of these allegations is a reflection upon the Son of God and is blasphemous (1989, pp. 25-26, parenthetical comment in orig.).

Furthermore, Paul affirmed the following:

For the invisible things of him **since the creation of the world** are clearly seen, being perceived through the things that are made, even his everlasting power and divinity; that they may be without excuse (Romans 1:20, emp. added).

The apostle declared that **from the creation of the world** the invisible things of God have been: (a) clearly seen; and (b) perceived. The phrase, "since the creation of the world," is translated from the Greek, *apo ktiseos kosmou*. As a preposition, *apo* is used "to denote the point from which something begins" (Arndt and Gingrich, 1957, p. 86). The term "world" is from the Greek, *kosmos*, and refers to "the orderly universe" (Arndt and Gingrich, p. 446). R.C. Trench observed that the *kosmos* is "the material universe...in which man lives and moves, which exists for him and of which he constitutes the moral centre" (1890, pp. 215-216). The term "perceived" is translated from the Greek *noeo*, which is used to describe rational, human intelligence. The phrase, "clearly seen" is an intensified form of *horao*, a word that "gives prominence to the discerning mind" (Thayer, 1962, p. 452). Both "perceived" and "clearly seen" are present tense forms, and as such denote "the continued manifestation of the being and perfections of God, by the works of creation from the beginning" (MacKnight, 1960, p. 58).

Who observed and perceived the things that were made "from the beginning" of the creation? If no man existed on this planet for billions of years (because man is a "relative newcomer to the Earth"), who was observing—with rational, human intelligence—these phenomena? Paul undoubtedly was teaching that **man** has existed since the creation of the world and has possessed the capacity to comprehend the truth regarding the existence of the Creator; accordingly, those who refuse to glorify Him are without excuse. It likewise is inexcusable for one who professes to believe the Bible as God's inspired Word to ignore such verses as these —or to twist and wrest them to try to make them say something they never were intended to say—merely to defer instead to evolutionary geology in an attempt to defend the concept of an ancient Earth. Yet examples of that very thing are all too prevalent.

During the question and answer session that followed my public debate with Jack Wood Sears on the topic of the Bible and the age of the Earth (see chapter 2), a querist asked him how he could defend the concept of an ancient Earth in light of Christ's statements in Mark 10:6 and Matthew 19:4 which indicated that "from the beginning of the creation, male and female made he them." Astonishingly, Dr. Sears responded by suggesting that neither Mark 10:6 nor Matthew 19:4 was addressing the creation of the **world**. Rather, he insisted, both passages meant "from the time of the creation **of man and woman**." What?! Were that the case, these two passages then would have the Lord saying, "From the beginning of the creation (of man and woman), man and woman created he them." The Son of God was not in the habit of talking in such nonsensical terms. Furthermore, Mark plainly wrote about "the beginning of **the** creation," not "**their** creation." Christ's point is crystal clear, especially when connected to Paul's comment in Romans 1:20-21 that someone with rational, **human** intelligence was "perceiving" the things that had been created. Riegle was right when he suggested: "It is amazing that men will accept long, complicated, imaginative theories and reject the truth given to Moses by the Creator Himself" (1962, p. 24).

"FROM THE BLOOD OF ABEL"

In Luke 11:45-52, the account is recorded of the Lord rebuking the rebellious Jews of His day. He charged them with following in the footsteps of their ancestors. He foretold the horrible destruction that was yet to befall them. And finally, He announced that upon them would come "the blood of all the prophets, which was shed **from the foundation of the world.**" Then, with emphatic linguistic parallelism (which so often is characteristic of Hebrew expression), He added, "**from the blood of Abel** unto the blood of Zachariah...."

Jesus therefore placed the murder of Abel near the "foundation of the world." Granted, Abel's death occurred some years after the initial creation, but it was close enough to that event for Jesus to state that it was associated with "the foundation of the world." If the world came into existence several billion years before the first family, how could the shedding of human blood be declared by God's Son to extend back to the "foundation of the world"?

Those who opt for an old-Earth scenario believe, of course, that man is a "recent addition" to the Earth—a "johnny-come-lately" who has been here only 3 million years or so out of an alleged Earth history of 4.6 **billion** years. It is apparent, however, that they are not obtaining their information from the same divine source as the prophet Isaiah who asked the skeptics of his day, "Hath it not been told you **from the beginning**? Have ye not understood **from the foundations of the earth**?" (40:21, emp. added). Isaiah understood that man had been on the Earth "from the beginning" or, as he stressed, "from the foundations of the Earth." Sad, is it not, that so many today who claim to believe the Bible refuse to acknowledge that simple, scriptural fact?

HOW LONG WERE ADAM AND EVE IN THE GARDEN OF EDEN?

On occasion, those who defend an old Earth suggest that it is impossible to know how long Adam and Eve were in the Garden of Eden and that untold years may have elapsed during that time period. Consider two popular arguments that frequently are offered in support of such a theory.

First, John Clayton has suggested that since a part of God's curse on Eve was that He was going to **multiply** her pain in childbirth (Genesis 3:16), she **must** have given birth to numerous children in the garden or God's curse would have meant nothing to her. How could God "multiply" something if she never had experienced it in the first place? Furthermore, Clayton has lamented, rearing children is a process that requires considerable time, thereby allowing for the possibility that Adam and Eve were in the Garden of Eden for an extended period prior to being evicted after their sin. As Clayton has written: "Every evidence we have biblically indicates that mankind's beginning in the Garden of Eden **was not a short period** which involved one man and one woman" (1980a, p. 5, emp. added).

The second argument (somewhat related to the first) suggests that Adam and Eve **must** have been in the garden for quite some time because after they left, it was said of Cain that "he builded a city" (Genesis 4:17). To quote Clayton, that is something "which you cannot do with you and your wife" (1980a, p. 5). In other words, Cain had to have a large enough family to help him build "a city." That, suggests Clayton, would have taken a lot of time.

Mr. Clayton is completely in error when he says that "every evidence we have biblically indicates that mankind's beginning in the Garden of Eden was not a short period which involved one man and one woman." The fact is, every evidence we have biblically proves conclusively that man and woman **could not have been in the garden for very long**. Consider the following.

First, regardless of what defenders of an ancient Earth may **wish** were, true, the simple fact of the matter is that the Bible sets an sets an outer limit on the amount of time that man could have lived in the Garden of Eden. Genesis 5:5 states clearly that "**all the days** that Adam lived were **930 years**." We know, of course, that "days" and "years" already were being counted by the time of Adam's creation because in Genesis 1:14 (day four of creation) God mentioned both in His discussion of their relationship to the heavenly bodies. Therefore, however long Adam and Eve may have been in the garden, one thing is for sure: they were not there for a time period that exceeded Adam's life span (930 years). Additionally, a certain portion of man's life was spent **outside** the Garden of Eden due to his sin against God—thereby reducing even further the portion of the 930 years that could have been spent in the garden setting.

Second, surely it is not inconsequential that **all** the children of Adam and Eve mentioned in the Bible were born **outside** the Garden of Eden. **Not one conception, or birth,** is mentioned as having occurred while Adam and Eve lived in the garden (see Genesis 4:1 for the first mention of any conception or birth—only **after** the couple's expulsion from Eden). Follow closely the importance and logic of this argument, which may be stated as follows.

One of the commands given to Adam and Eve was that they "be fruitful and multiply, and fill the Earth" (Genesis 1:28). [Interestingly, Isaiah would say many years later that God created the Earth "to be inhabited" (Isaiah 45:18).] In other words, Adam and Eve were to **reproduce**.

But what is sin? Sin is: (a) **doing** what God said **not to do**; or (b) **not doing** what God said **to do**. Up until the time that Adam and Eve ate the fruit of the tree of the knowledge of good and evil (Genesis 3:6), had they sinned? No, they still were in a covenant relationship with God and everything was perfect. Since that is the case, the only conclusion that can be drawn is that Adam and Eve were doing what God had commanded them to do—reproducing. Yet, I repeat, the only conceptions and births of which we have any record occurred **outside the garden**! In other words, apparently Adam and Eve were not even in the garden long enough for Eve to conceive, much less give birth.

Third, while the Bible does not provide a **specific** time regarding how long Adam and Eve were in the Garden, it could not have been very long because Christ Himself, referring to the curse of death upon the human family as a result of their sinful rebellion against God, specifically stated that the devil "was a murderer **from the beginning**" (John 8:44). Satan and his ignominious band of outlaws ("sons of the evil one"—Matthew 13:38) have worked their ruthless quackery on mankind from the very moment the serpent met mother Eve in the Garden of Eden. When he and his cohorts rebelled and "kept not their proper habitation," they were cast from the heavenly portals to be "kept in everlasting bonds under darkness unto the judgment of the great day" (Jude 6).

Satan fought with God—and lost. The devil's insurrection had failed miserably, and that failure had dire, eternal consequences. His obstinate attempt to usurp God's authority cost him his position among the heavenly hosts. As a result of his rebellion, he was cast "down to hell" (2 Peter 2:4). In the end, his sedition gained him nothing and cost him everything. Regardless of the battle plan he adopted to challenge the Creator of the Universe, regardless of the battlefield he chose as his theater of war, and regardless of the strength or numbers of his army, the simple fact of the matter is that—in the most important contest of his existence—he lost! The conditions of his ultimate surrender were harsh. Although his armies had been thoroughly routed, although he had been completely vanquished, and although the Victor had imposed the worst kind of permanent exile, Satan was determined not to go gently into the night. While he had lost the war, he nevertheless planned future skirmishes. Vindictive by nature (Revelation 12:12), in possession of cunning devices (2 Corinthians 2:11), and determined to be "the deceiver of the world" (Revelation 12:9), he set his face against all that is righteous and holy—and never looked back. His anger at having been defeated fueled his determination to strike back in revenge.

But strike back at whom? God's power was too great, and His omnipotence was too all-consuming (Job 42:2; 1 John 4:4). Another target was needed; another repository of satanic revenge would have to be found. And who better to serve as the recipient of hell's unrighteous indignation than mankind—the only creature in the Universe made "in the image and likeness of God" (Genesis 1:26-27)? As Rex A. Turner, Sr. has observed: "Satan cannot attack God directly, thus he employs various methods to attack man, God's master creation" (1980, p. 89). What sweet revenge—despoiling the "apple of God's eye" and the zenith of His creative genius! Thus, with the creation of man, the battle was on. Little wonder, then, that in his first epistle the apostle Peter described Satan as an adversary who, "as a roaring lion, walketh about, seeking whom he may devour" (5:8).

Now—knowing what we the Scriptures tell us about Satan's origin, attitude, and mission—is it sensible to suggest that he would take his proverbial time, and twiddle his figurative thumbs, while allowing Adam and Eve to revel in the covenant relationship they enjoyed with their Maker (read Genesis 3:8 about how God walked with them in the garden "in the cool of the day")? Would he simply "leave them alone for a long period of time" so that they could conceive, give birth to, and rear children in the luscious paradise known as the Garden of Eden? Is this how a hungry, stalking lion would view its prey—by watching admiringly from afar, allowing it hundreds or thousands of years of fulfilled joy, and affording it time to conceive, give birth to, and raise a family? Hardly—which is why Christ described Satan as a murderer "from the beginning." Satan was in no mood to wait. He was angry, he was bitter, and he was filled with a thirst for revenge. What better way to slake that thirst than introducing sin into God's perfect world?

What may be said, then, about John Clayton's suggestion that Adam and Eve must have been in the Garden for an extended time period because God said that He was going to "multiply"

Eve's pain. How could he "multiply" something she never had experienced? This quibble can be answered quite easily. Does a person have to "experience" something before that something can be "multiplied"? Suppose I said, "I'm going to give you $100." You, therefore, eagerly stick out your hand to receive the $100 bill I am holding in mine. But, as you reach, I immediately pull back my hand and say, "No, I've changed my mind; I'm going to give you $1,000 instead!" Did you actually have to possess or "experience" the $100 bill before I could increase it to $1,000? Of course not!

The fact that God said He intended to "multiply" Eve's pain in childbirth does not mean necessarily that Eve had to have experienced **some** pain before God's decree that she would experience **more** pain. God's point was merely this: "Eve, you were going to experience pain in childbirth, but because of your sin, now you will experience even more pain." The fact that Eve never had experienced **any** childbirth pain up to that point does not mean that she could not experience **even more** pain later as a part of her penalty for having sinned against God.

Lastly, what about John Clayton's idea that Adam and Eve must have been in the Garden for an extended period of time because the text indicates that when they left Cain and his wife "builded a city" (Genesis 4:17). Clayton has lamented that this is something "which you **cannot do with you and your wife**" (1980a, p. 5). Of course, he would be correct—**if** the city under discussion were a modern metroplex. But that is not the case here.

The Hebrew word for city is quite broad in its meaning. It may refer to anything from a sprawling village to a mere encampment. Literally, the term means "place of look-out, especially as it was fortified." In commenting on Genesis 4:17, Old Testament commentator John Willis observed: "However, a 'city' is not necessarily a large, impressive metropolis, but may be a small unimposing village of relatively few inhabitants" (1979, p. 155). Again, apply some common sense here. What would it be **more likely** for the Bible to suggest that Cain and his wife constructed (considering who they were and where they were living)—a thriving, bustling, metropolis, or a Bedouin tent city. To ask is to answer, is it not? To this very day, Bedouin tent cities are commonplace in that particular area of the world. And, as everyone will admit, two boy scouts can erect a tent, so it does not strain credulity to suggest that likely Cain and his wife were able to accomplish such a task as well.

THE DOCTRINE OF APPARENT AGE

On occasion, the comment is overheard, "But the Earth **looks so old**." There are at least two responses that might be made to such a statement. First, one might ask, "Compared to what; what does a **young** Earth look like?" Who among us has anything with which to compare? Second, we should not be surprised if certain methods in science appear to support the idea of an ancient Earth. Why? The answer lies in what has been called the "doctrine of apparent age" (also known as the "doctrine of mature creation").

This concept states that when God made "heaven, and earth, the sea, and all that in them is" (Exodus 20:11), they were made perfect, complete, and ready for habitation by mankind and the multiple forms of plant and animal life. God did not create **immature** forms (although He certainly could have done so, had He wished), but **mature** ones. Rather than creating an acorn, for example, He created an oak. Rather than creating an egg, He created a chicken. Rather than creating Adam and Eve as infants or young children, He created them as post-pubescent beings. We know this to be true because one of the commands God gave each living thing shortly after its creation was that it should reproduce "after its kind." This very command, in fact, was given to Adam and Eve while they still were in the Garden of Eden, prior to their sin and expulsion.

How old were Adam and Eve two seconds after their creation? They were two seconds old. How old were the plants and animals two seconds after their creation? They were two seconds old. But how old all these two-second-old people, plants, and animals **look** like they were? Trevor Major has commented:

> So Adam, for example, had the look and the capability of a full-grown man on the first Sabbath, even though he had lived only one day. Thus, according to the doctrine of mature creation, all living things were created in a mature state, with only the **appearance** of age (1989, p. 16, emp. in orig.).

It is important to realize that the initial creation had **two** ages—a **literal** age, and an **apparent** age. It **literally** may have been just one day old, two days old, three days old, and so on. But it **appeared** to be much older.

The biblical record provides additional information concerning the accuracy of the doctrine of apparent age. In Genesis 1: 14, God told Moses that the heavenly bodies (e.g., Sun, Moon, stars) were to be "for signs and for seasons, for days and for years." In order for the heavenly bodies to be useful to man for the designation of signs, seasons, days, and years, those heavenly bodies **must have been visible**. Thus, when God created them He made their light already visible from the Earth. The psalmist exclaimed: "The heavens declare the glory of God, and the firmament showeth his handiwork" (19:1). There was, therefore, **purpose** behind God's mature creation.

First, the Earth was prepared in a mature state so that man would find it suitable for his habitation. Christ specifically stated that man and woman had been on the Earth "since the beginning of the creation" (Matthew 19:4; Mark 10:6). Thus, it was necessary that "from the beginning" the Earth be "finished." Second, once man found himself in such a home (called "very good"—denoting complete perfection—in Genesis 1:31), it was only right to give honor and glory to the Creator Who designed and built such a magnificent edifice. This explains why Paul, in Romans 1:20ff., suggested that even God's "everlasting power and divinity" had been seen by mankind "from the creation of the world," and why those who refused to honor God would be "without excuse."

Even the miracles of the Bible reflect God's frequent use of the principle we call "apparent age." During Christ's first miracle, He transformed water into wine (John 2).* For mere mortals to produce wine (alcoholic or not) requires a lengthy process employing soil, water, grapes, sunshine, etc. Yet Jesus accomplished this task in mere minutes, producing what the governor of the wedding feast termed not just wine but "good wine" (John 2:10). The miracle of the feeding of the 5,000 (Matthew 14:13-21) also provides evidence regarding the principle of apparent age. The young boy present on that occasion had but a few loaves and fishes, yet Christ "multiplied" them and fed over 5,000 men alone. Major addressed this concept when he wrote:

> Thousands of loaves were distributed for which the barley had not been sown, harvested, or milled, and which had never been mixed into dough and baked in an oven. Equally amazing, thousands of dried fishes were handed out which neither had grown from an egg nor been caught in a fishermen's net. Everything was there in a prepared form, ready to eat by the recipients of this great wonder.

> The miracle of creation was also achieved in a relative instant, producing an effect which could have only a supernatural cause. In the first chapter of Genesis, God created trees and grasses, not just their seeds. He created birds which could already fly, not eggs or even chicks. He created fish which could already swim, not fish eggs. He created cattle, not calves. And He created man and woman, not boy and girl. Speaking to these animals, and to these people, God commanded: "Be fruitful and multiply" (1:22,28). Notice that

---

* In the New Testament, the Greek term for wine, *oinos*, is employed to denote both alcoholic and non-alcoholic grape juice beverages, prohibiting the view that the wine spoken of in John 2 necessarily was alcoholic.

the plants and animals began to multiply according to their own kind almost straightaway (1:11,24). Immature organisms could not have reproduced, and in any case, would have perished in the absence of their adult forms (1989, 27 [10]:16, emp. in orig.).

The moment God created matter itself, would it not have appeared "mature"—i.e., as if it already had existed? If God had decided to create Adam as a baby, how could He have produced a baby that did not **look** like it had gone through a nine-month gestation period? If He had created an acorn, how could He have created an acorn that did not **look** like it had fallen from a mighty oak? Did God create the Earth "mature"? Indeed He did. How could He have done otherwise?

However, we must be careful not to abuse this concept. Some have asked if the Creator might have placed fossils (or, for that matter, fossil fuels) within the Earth in order to make it "appear" ancient. This idea should be rejected for several important reasons. First, such a suggestion implies that the formation of fossils and/or fossil fuels is an inherently slow, uniformitarian-type process—which it is not.

Second, certain geological/paleontological phenomena provide some of the best examples available in the world around us of a sudden, global, non-uniformitarian catastrophe (i.e., the Genesis Flood). When the doctrine of apparent age is invoked in an inappropriate manner, it robs mankind of powerful testimony to the workings of the Creator and weakens the similarly powerful testimony of His Word regarding what He did and how He did it.

Third, the idea that the Creator may have "planted" such things as fossils and fossil fuels in the Earth is an indictment of the nature and character of God, Who never would try to "trick" or "fool" man in such a way. Nor would He ever lie (Titus 1:2). If we observe things in the Earth like fossils, fossil fuels, etc., we naturally (and rightly) assume that these are the results of **real** plants and/or animals that actually lived. It will not do for us to say,

"God just put them there," for such a suggestion makes God deceptive, which He is not. As Major has stated: "Tactics of confusion and deception hardly belong to a Creator Who would have humanity discern Him by His creation (Romans 1:20)" [1989, 27[10]:16].

Others have suggested that if God created things that appear older than they really are, that is deceptive on the face of it. Thus, by definition the doctrine of apparent age makes God out to be a liar and should be rejected on that count alone. However, such an accusation overlooks the fact that **God plainly told us what He did!** Anyone who takes the time to read Genesis 1-2 can see within those chapters God's methodology. In fact, He made certain to tell us exactly how the Earth and its inhabitants came into existence. Perhaps—just perhaps—if God had not told us what He did, or had not been as specific as He was, **then** He might be accused of deception or trickery. But no one can accuse God (justifiably) of such despicable behavior because His Word explains His actions. He did not hide the facts from us but, quite the contrary, went to great lengths to reveal them.

Some have suggested that one of the most difficult questions relating to the doctrine of the apparent age has to do with the starlight that is seen by those of us on the Earth. The argument usually goes something like this. We know today that light travels at a speed slightly in excess of 186,000 miles per second. The time it takes light to travel one year is referred to as a light-year. Yet we are able to see light from stars that are millions of light-years away. How can this be if the Earth is young (with an age measured in thousands, not millions or billions of years)? Of course, a partial answer lies in the fact that God created the light from heavenly bodies already en route and visible to the Earth's inhabitants (as a part of the mature creation). Without that light, the night sky would lack patterns necessary for the signs, seasons, days, and years specified so clearly in Genesis 1:14, and mankind would not have been able to see God's "glory and handiwork" (Psalm 19:1).

Other issues may be involved as well, a discussion of which (e.g., the possibility that the speed of light has diminished over time, etc.) has been provided by various writers (see: Norman and Setterfield, 1987; Major, 1987; Ex Nihilo, 1984; Humphreys, 1994). The reader interested in an examination of these matters is referred to these sources and others that they may recommend. Such a discussion is beyond the purview of this book, however, since it has to do more with scientific matters than biblical.

The doctrine of apparent age not only explains many of the alleged evidences for an ancient Earth, but is entirely scriptural in its foundations. It helps answer many of the questions relating to data that evolutionists, and those sympathetic to them, offer as documentation for their concept of a planet of great antiquity —which, in reality, is one of relative youth.

7

CONCLUSION

There are many people who accept unreservedly the Bible's teaching on matters of both a spiritual and a physical nature. They do not believe in evolution, and defend as genuine the Bible's instruction regarding such topics as its own inspiration, Christ's deity, and the importance of the church. They acknowledge that God "has granted unto us all things that pertain unto life and godliness" (2 Peter 1:3). And, to the very best of their ability, they live "soberly, righteously, and godly in this present world" (Titus 2:12). Yet when it comes to the Bible's teaching on the age of the Earth, they simply shrug their shoulders (as if they do not quite know what to do with the information) and are content to take a somewhat "agnostic" stance in regard to the biblical information on this crucial topic. Apparently, they are undecided about what to do with the Bible's teachings in this area, especially since "science" seems to be offering them a conclusion diametrically opposed to the one dictated by the Bible. In the end, for whatever reason(s), "science" wins as they set aside biblical instruction in favor of current scientific theory.

But why is this the case? The Bible **does** address the topic of the age of the Earth, as our discussion here amply documents. If a person is willing to accept the Bible's instructions on its own inspiration, God's existence, Christ's deity, the need to live a decent, honest, moral life, and hundreds of other topics that deal not just with godliness but with "**life** and godliness," why, then, can that same person not accept the Bible's simple, straightforward teaching on the age of the Earth? Is one set of instructions any more difficult to believe than the other? Our plea is for such Bible believers to be consistent and to abandon the concept of an ancient Earth that is so foreign to the Scriptures. Accept **all** that the Bible has to say—including its plain statements and clear implications regarding the age of the Earth.

No doubt there also are many Bible believers who simply do not know **what** to do regarding the problem of the age of the Earth. They "lean" toward belief in an old Earth, but only because they never have stopped to consider that **one of the most compelling reasons for belief in an old Earth is to legitimize the concept of evolution** (without an ancient planet, evolution obviously is impossible). But were someone to ask, "Do you believe in evolution?," their answer likely would be, "No, I do not." Then why believe in an old Earth? Why not simply examine what the Bible says regarding the age of the Earth and accept it forthwith? On occasion, the person who starts out conceding an ancient Earth eventually ends up in the evolutionists' camp. At some point, he or she is led to think: If the Earth really is billions of years old, then perhaps evolution has been going on for all that time after all.

How old is the Earth? Biblically speaking, it is **five days older than man**! Relatively speaking, it is quite young—with an age measured in thousands, not billions, of years. Some, of course, have ridiculed such an idea. Rubel Shelly, in his book, *What Shall We Do with the Bible?* wrote:

> Notice an example of a case where an ignorance of the Bible creates an apparent conflict between it and science. There are countless individuals who honestly believe that the cre-

ation of the world took place in 4004 B.C. **Yet the evidence available to scientists points to a very old earth, possibly several billion years old.... We just do not know when the creation of this planet occurred!** (1975, pp. 39-40, emp. added).

Fifteen years later, in his book, *Prepare to Answer: A Defense of the Christian Faith*, Dr. Shelly went even further when he suggested: "Few would argue that the earth is 'only about 6,000 years old'...," and that such a position is "not held by responsible apologists" (1990, p. 61).

I suggest, however, that such a position **is** held by "responsible apologists," because the Bible is factual in its clear statements and implied deductions regarding the Earth and man's history on it. Christians should not be stampeded into accepting the compromising views of evolutionary scientists—or those sympathetic to them. Let us not be destroyed "for lack of knowledge" (Hosea 4:6).

REFERENCES

Ackerman, Paul (1986), *It's a Young World After All* (Grand Rapids, MI: Baker).

Aling, C. (1981), *Egypt and Bible History* (Grand Rapids, MI Baker).

Archer, Gleason L. (1970), *Old Testament Introduction* (Chicago, IL: Moody).

Archer, Gleason L. (1979), "The Chronology of the Old Testament," *The Expositor's Bible Commentary* (Grand Rapids, MI: Eerdmans).

Archer, Gleason L. (1982), *Encyclopedia of Bible Difficulties* (Grand Rapids, MI: Zondervan).

Arndt, William and F.W. Gingrich (1957), *A Greek-English Lexicon of the New Testament and Other Early Christian Literature* (Chicago, IL: University of Chicago Press).

Baikie, James (1929), *A History of Egypt* (London: A&C Black).

Bloomfield, S.T. (1837), *The Greek New Testament with English Notes* (Boston, MA: Perkins and Marvin).

Brandfon, Fredric (1988), "Archaeology and the Biblical Text," *Biblical Archaeology Review*, 14[1]:54-59, January/February.

Brantley, Garry K. (1993), "Dating in Archaeology: Challenges to Biblical Credibility," *Reason & Revelation*, 13:82-85, November.

Breasted, James (1912), *History of Egypt* (New York, Charles Scribner's Sons).

Brown, Francis et al. (1979), *The New Brown-Driver-Briggs-Gesenius Hebrew and English Lexicon* (Peabody, MA: Hendrickson).

Carnell, Edward John (1959), *The Case for Orthodox Theology* (Philadelphia, PA: Westminster Press).

Cassel, J. Frank (1973), "Biology," *Christ and the Modern Mind*, ed. Robert W. Smith (Downers Grove, IL: InterVarsity Press).

Clayton, John N. (no date-a), "Biblical Misconceptions and the Theory of Evolution," *Does God Exist? Correspondence Course*, Lesson 4.

Clayton, John N. (no date-b), "The History of Man on Planet Earth," *Does God Exist? Correspondence Course*, Lesson 8.

Clayton, John N. (no date-c), *Questions and Answers: Number One* (South Bend, IN: privately published by author), taped lecture.

Clayton, John N. (no date-d), *Evolution's Proof of God* (South Bend, IN: privately published by author), taped lecture.

Clayton, John N. (1976a), *The Source* (South Bend, IN: privately published by author).

Clayton, John N. (1976b), "'Flat Earth' Bible Study Techniques," *Does God Exist?*, 3[10]:2-7, October.

Clayton, John N. (1977), "The 'Non-World View' of Genesis," *Does God Exist?*, 4[6]:6-8, June.

Clayton, John N. (1979), "The Necessity of Creation—Biblically and Scientifically," *Does God Exist?*, 6[5]:2-5, May.

Clayton, John N. (1980a), "Is the Age of the Earth Related to a 'Literal Interpretation' of Genesis?," *Does God Exist?*, 7[1]:3-8, January.

Clayton, John N. (1980b), *A Response to "Evolutionary Creationism"* (South Bend, IN: privately published by author), taped lecture.

Clayton, John N. (1982), "Where Are the Dinosaurs?," *Does God Exist?*, 9[10]:2-6, October.

Clayton, John N. (1989), "How Much Does Modernism Rob Us of Biblical Understanding?," *Does God Exist?*, 16[1]:4-7, January/February.

Clayton, John N. (1990a), "One Week Creation—Of Man or of God?," *Does God Exist?*, 17[4]:5-12, July/August.

Clayton, John N. (1990b), *The Source* (South Bend, IN: privately published by author).

Clayton, John N. (1990c), "The History of the Earth," *Does God Exist? Correspondence Course*, Lesson 9.

Clayton, John N. (1990d), "How Did God Create Man?," *Does God Exist? Correspondence Course*, Lesson 7.

Clayton, John N. (1991), "Creation Versus Making—A Key to Genesis 1," *Does God Exist?*, 18[1]:6-10, January/February.

Coffman, Burton (1985), *Commentary on Genesis* (Abilene, TX: ACU Press).

Cremer, H. (1962), *Biblico-Theological Dictionary of New Testament Greek* (London: T&T Clark).

Culp, G. Richard (1975), *Remember Thy Creator* (Grand Rapids, MI: Baker).

Custance, Arthur (1967), *The Genealogies of the Bible*, Doorway Paper #24 (Ottawa, Canada: Doorway Papers).

Custance, Arthur (1970), *Without Form and Void* (Brockville, Canada: Doorway Papers).

DeHoff, George (1944), *Why We Believe the Bible* (Murfreesboro, TN: DeHoff Publications).

Dods, Marcus (1948), "Genesis," *The Expositor's Bible*, ed. W.R. Nicoll (Grand Rapids, MI: Eerdmans).

England, Donald (1972), *A Christian View of Origins* (Grand Rapids, MI: Baker).

England, Donald (1982), Letter to Dr. Clifton L. Ganus, President, Harding University, Searcy, Arkansas.

England, Donald (1983), *A Scientist Examines Faith and Evidence* (Delight, AR: Gospel Light).

Ex Nihilo (1984), "Update," 6[4]:46.

Fields, Weston W. (1976), *Unformed and Unfilled* (Grand Rapids, MI: Baker).

Geisler, Norman L. and Ronald M. Brooks (1990), *When Skeptics Ask* (Wheaton, IL: Victor).

Gould, Stephen J., ed. (1993), *The Book of Life* (New York: W.W. Norton).

Green, William H. (1890), "Primeval Chronology," *Bibliotheca Sacra*, 47:294-295.

Harris, R.L., G.L. Archer, Jr., and B.K. Waltke, eds. (1980), *Theological Wordbook of the Old Testament* (Chicago, IL: Moody).

Hayward, Alan (1985), *Creation and Evolution: The Facts and the Fallacies* (London: Triangle Books).

Henkel, M. (1950), "Fundamental Christianity and Evolution," *Modern Science and the Christian Faith*, ed. F. Alton Everest (Wheaton, IL: Van Kampen Press).

Humphreys, D. Russell (1994), *Starlight and Time* (Green Forest, AR: Master Books).

Jackson, Wayne (1974), *Fortify Your Faith in an Age of Doubt* (Stockton, CA: Courier Publications).

Jackson, Wayne (1978), "The Antiquity of Human History," *Words of Truth*, 14[18]:1, April 14.

Jackson, Wayne (1981), "The Chronology of the Old Testament in the Light of Archaeology," *Reason & Revelation*, 1:37-39, October.

Jackson, Wayne (1989), *Creation, Evolution, and the Age of the Earth* (Stockton, CA: Courier Publications).

Jackson, Wayne (1990), "The Saga of Ancient Jericho," *Reason & Revelation*, 10:17-19, April.

Jackson, Wayne and Bert Thompson (1992), *In the Shadow of Darwin: A Review of the Teachings of John N. Clayton* (Montgomery, AL: Apologetics Press).

Jacobus, Melancthon (1864), *Notes on Genesis* (Philadelphia, PA: Presbyterian Board of Publication).

Jamieson, Robert, et al. (1945), *Jamieson, Faucett, Brown Bible Commentary* (Grand Rapids, MI: Eerdmans).

Jordan, James (1979/1980), "The Biblical Chronology Question," *Creation Social Sciences and Humanities Quarterly*, Winter, 1979, 2[2]:9-15; Spring, 1980, 2[3]:17-26.

Kantzer, Kenneth (1982), "Guideposts for the Current Debate over Origins." *Christianity Today*, pp. 23-25, October 8.

Kautz, Darrel (1988), *The Origin of Living Things* (Milwaukee, WI: privately published by author).

Keil, C.F. (1971 reprint), *The Pentateuch* (Grand Rapids, MI: Eerdmans).

Kitchen, Kenneth A. and J.D. Douglas, eds. (1982), *The New Bible Dictionary* (Wheaton, IL: Tyndale), second edition.

Klingman, George (1929), *God Is* (Cincinnati, OH: F.L. Rowe).

Klotz, John (1955), *Genes, Genesis, and Evolution* (St. Louis, MO: Concordia).

Laughlin, John (1992), "How to Date a Cooking Pot," *Biblical Archaeology Review*, 18[5]:72-74.

Lenski, R.C.H. (1961), *The Interpretation of St. Matthew's Gospel* (Minneapolis, MN: Augsburg).

Leupold, H.C. (1942), *Expositions of Genesis,* (Grand Rapids, MI: Baker, reprint).

MacKnight, James (1960 reprint), *Apostolical Epistles* (Nashville, TN: Gospel Advocate).

Major, Trevor J. (1987), "Questions and Answers," *Reason & Revelation*, 7:5-7, February.

Major, Trevor J. (1989), "Which Came First—The Chicken or the Egg?," *Bible-Science Newsletter*, 27[10]:16, October.

Major, Trevor J. (1993), "Dating in Archaeology: Radiocarbon and Tree-Ring Dating," *Reason & Revelation*, 13:74-77, October.

McClish, Dub, ed. (1983), *Studies in Hebrews* (Denton, TX: Pearl Street Church of Christ).

McIver, Tom (1988), "Formless and Void: Gap Theory Creationism," *Creation/Evolution*, 8[3]:1-24, Fall.

Merrill, E.H., (1978), *An Historical Survey of the Old Testament* (Phillipsburg, NJ: Presbyterian and Reformed).

Milligan, Robert (1972 reprint), *The Scheme of Redemption* (Nashville, TN: Gospel Advocate).

Morris, Henry M. (1966), *Studies in the Bible and Science* (Grand Rapids, MI: Baker).

Morris, Henry M. (1974), *Scientific Creationism* (San Diego, CA: Creation-Life Publishers).

Morris, Henry M. (1976), *The Genesis Record* (Grand Rapids, MI: Baker).

Morris, Henry M. (1984), *The Biblical Basis for Modern Science* (Grand Rapids, MI: Baker).

Morris, Henry M. (1989), *The Long War Against God* (Grand Rapids, MI: Baker).

Morris, Henry M. and Gary E. Parker (1987), *What Is Creation Science?* (San Diego, CA: Master Books, second edition).

Morris, John D. (1994), *The Young Earth* (Green Forest, AR: Master Books).

Norman, Trevor and Barry Setterfield (1987), *The Atomic Constants, Light, and Time*, Technical Report (Menlo Park, CA: Stanford Research Institute International).

Packer, J.I., Merrill C. Tenney, and William White, Jr. (1980), *The Bible Almanac* (Nashville, TN: Nelson).

Pember, George H. (1876), *Earth's Earliest Ages* (New York: Revell).

Pilgrim, James (1976), "Day Seven," *Gospel Advocate*, 118[33]:522, August 12.

Ramm, Bernard (1954), *The Christian View of Science and Scripture* (Grand Rapids, MI: Eerdmans).

Rehwinkle, Alfred (1974), *The Wonders of Creation* (Grand Rapids, MI: Baker).

Riegle, D.D. (1962), *Creation or Evolution?* (Grand Rapids, MI: Zondervan).

Rimmer, Harry (1937), *Modern Science and the Genesis Record* (Grand Rapids, MI: Eerdmans).

Ross, Hugh (1991), *The Fingerprint of God* (Orange, CA: Promise Publishing Co.).

Ross, Hugh (1994), *Creation and Time* (Colorado Springs, CO: Navpress).

Schroeder, Gerald L. (1990), *Genesis and the Big Bang* (New York: Bantam).

Schroeder, Gerald L. (1997), *The Science of God* (New York: Free Press).

Scofield, Cyrus I., ed. (1917), *Scofield Reference Bible* (New York: Oxford University Press).

Sears, Jack Wood (1969), *Conflict and Harmony in Science and the Bible* (Grand Rapids, MI: Baker).

Shelly, Rubel (1975), *What Shall We Do with the Bible?* (Jonesboro, AR: National Christian Press).

Shelly, Rubel (1990), *Prepare to Answer: A Defense of the Christian Faith* (Grand Rapids, MI: Baker).

Smith, Wilbur M. (1945), *Therefore Stand!* (Grand Rapids, MI: Baker).

Surburg, Raymond (1959), "In the Beginning God Created," *Darwin, Evolution, and Creation*, ed. P.A. Zimmerman (St. Louis, MO: Concordia).

Taylor, Ian (1984), *In the Minds of Men* (Toronto: TFE Publishing).

Taylor, Kenneth (1974), *Evolution and the High School Student* (Wheaton, IL: Tyndale).

Thayer, J.H. (1962), *Greek-English Lexicon of the New Testament* (Grand Rapids, MI: Zondervan).

Thiele, Edwin (1977), *A Chronology of the Hebrew Kings* (Grand Rapids, MI: Zondervan).

Thomas, J.D. (1961), *Evolution and Antiquity* (Abilene, TX: Biblical Research Press).

Thompson, Bert (1977), *Theistic Evolution* (Shreveport, LA: Lambert).

Thompson, Bert (1982), "The Gap Theory: Still Another False Compromise of the Genesis Account of Creation," *Reason & Revelation*, 2:45-48, November.

Thompson, Bert (1995), *Creation Compromises* (Montgomery, AL: Apologetics Press).

Thompson, Bert (1998a), "Satan—His Origin and Mission [Part I]," *Reason & Revelation*, 18:73-79, October.

Thompson, Bert (1998a), "Satan—His Origin and Mission [Part II]," *Reason & Revelation*, 18:81-87, November.

Thurman, Clem (1986), "The Genesis 'Day' and Age of Earth," *Gospel Minutes*, 35[14]:2-3.

Trench, R.C. (1890), *Synonyms of the New Testament* (London: Kegan, Purl, Trench, Trubner & Co.).

Turner, Rex A., Sr. (1980), *Systematic Theology* (Montgomery, AL: Alabama Christian School of Religion).

Unger, Merrill (1966), *Unger's Bible Handbook* (Chicago, IL: Moody).

Unger, Merrill (1973), *Archaeology and the Old Testament* (Grand Rapids, MI, Zondervan).

Van Bebber, Mark and Paul S. Taylor (1996), *Creation and Time: A Report on the Progressive Creationist Book by Hugh Ross* (Gilbert, AZ: Eden Communications).

Van Till, Howard J. (1986), *The Fourth Day* (Grand Rapids, MI: Zondervan).

Van Till, Howard J., R.E. Snow, J.H. Stek, and Davis A. Young (1990), *Portraits of Creation* (Grand Rapids, MI: Eerdmans).

Whitcomb, John C. (1972), *The Early Earth* (Grand Rapids, MI: Baker).

Whitcomb, John C. (1973a), "The Days of Creation," *And God Created*, ed. Kelly L. Segraves (San Diego, CA: Creation-Science Research Center), 2:61-65.

Whitcomb, John C. (1973b), "The Gap Theory," *And God Created*, ed. Kelly L. Segraves (San Diego, CA: Creation-Science Research Center), 2:67-71.

White, J.E.M. (1970), *Ancient Egypt* (New York: Dover).

Wilder-Smith, A.E. (1975), *Man's Origin: Man's Destiny* (Minneapolis, MN: Bethany Fellowship).

Williams, Arthur F. (1965), *Creation Research Annual* (Ann Arbor, MI: Creation Research Society).

Williams, Arthur F. (1970), "The Genesis Account of Creation," *Why Not Creation?*, ed. Walter E. Lammerts (Grand Rapids, MI: Baker).

Willis, John T. (1979), "Genesis," *The Living Word Commentary* (Austin, TX: Sweet).

Woodmorappe, John (1999), *The Mythology of Modern Dating Methods* (El Cajon, CA: Institute for Creation Research).

Woods, Guy N. (1976), *Questions and Answers: Open Forum* (Henderson, TN: Freed-Hardeman University).

Wysong, R.L. (1976), *The Creation-Evolution Controversy* (East Lansing, MI: Inquiry Press).

Young, Davis A. (1977), *Creation and the Flood* (Grand Rapids, MI: Baker).

Young, Edward J. (1964), *Studies in Genesis One* (Nutley, NJ: Presbyterian and Reformed)